HARNESSING COMPLEXITY

Organizational Implications of a Scientific Frontier

Robert Axelrod
and
Michael D. Cohen

The Free Press
New York • London •
Toronto • Sydney • Singapore

THE FREE PRESS
A Division of Simon & Schuster Inc.
1230 Avenue of the Americas
New York, NY 10020

THE FREE PRESS and colophon are trademarks
of Simon & Schuster Inc.

Designed by Deirdre C. Amthor

Manufactured in the United States of America

10   9   8   7   6   5   4   3   2   1

Library of Congress Cataloging-in-Publication Data
Axelrod, Robert M.
  Harnessing complexity : organizational implications of a scientific frontier /
Robert Axelrod and Michael D. Cohen.
    p. cm.
  Includes bibliographical references and index.
    1. Complex organizations. 2. System analysis. I. Cohen, Michael D. II. Title.

HM131 .A897 2000
302.3'5–dc21   99-058063

ISBN 0-684-86717-6

To the BACH Group

Contents

Preface

This is a small book about a large question: In a world where many players are all adapting to each other and where the emerging future is extremely hard to predict, what actions should you take?

We call such worlds Complex Adaptive Systems. In Complex Adaptive Systems there are often many participants, perhaps even many kinds of participants. They interact in intricate ways that continually reshape their collective future. New ways of doing things—even new kinds of participants—may arise, and old ways—or old participants—may vanish. Such systems challenge understanding as well as prediction. These difficulties are familiar to anyone who has seen small changes unleash major consequences. Conversely, they are familiar to anyone who has been surprised when large changes in policies or tools produce no long-run change in people's behavior.

When managers and policy makers hear about complexity research, they often ask, "How can I control complexity?" What they usually mean is, "How can I eliminate it?" But complexity, as we shall see, stems from fundamental causes that cannot always be eliminated. Although complexity is often perceived as a liability, it can actually be an asset. The thesis of this book is that complexity can be harnessed. So, rather than seeking to eliminate complexity, we explore how the dynamism of a Complex Adaptive System can be used for productive ends. Therefore, we ask how organizations and strategies can be designed to take advantage of the opportunities provided by complexity.

In a world of mutually adaptive players, even though prediction may be difficult, there is quite a bit that you can do. Complexity itself allows for techniques that promote effective adaptation. When there are many participants, numerous interactions, much trial-and-error learning, and abundant attempts to imitate each other's successes, there will also be rich opportunities to harness the resulting complexity. And there will be things to avoid. To take a simple example: Even though one action seems best, it usually pays to maintain variety among the actions you take so that you can continue to learn and adapt. The purpose of this book is to help managers and policy makers harness complexity.

We address a variety of readers. Some of our readers may simply be interested in learning more about the exciting research frontier called complexity. For these readers we provide a nontechnical introduction to the field. Other readers may have more specific interests in how to design social systems, or in how to make better policy for existing social systems. For these readers we draw on a wide range of applications from business, political, and cultural settings. We make no assumptions about the backgrounds of our readers, except that they are curious about how social systems work and how they can be made to work better.

As we were developing the organizational implications of complexity, we saw the need to bring order to the vast range of research in the field. To do this, we constructed a framework that provides a systematic way to analyze a particular setting and thereby suggest useful questions and promising possibilities for action. We found that our framework helped to clarify some deep relationships among many hitherto separate lines of complexity research. Moreover, the framework uncovered some important gaps. To fill them, we made a number of specific contributions to complexity research. These include the critical role of nonrandom interactions in adaptation, the contrast of biological with informational copying, and the relationships between credit allocation and measures of performance.

The foundations of this book lie in three distinct fields: evolutionary biology, computer science, and social design.

From evolutionary biology come the insights of Darwinian evolution, particularly that extraordinary adaptations can come about through the selection and reproduction of successful individuals in populations. Even though moths in England could not understand or predict that the Industrial Revolution would turn white-barked trees into soot-covered trees, it did not take very long for selection by predatory birds to transform the population of moths near a factory from white to black.

From computer science come insights about how systems with many artificial agents can be designed to work together and even adapt over time to each other and to their ever-changing environment. Two areas of computer science have been especially important to us. First, there is the field of evolutionary computation, which has fostered an engineering approach to adaptation. With an engineering approach, one asks how systems can be designed to become more effective over time. By making evolution and adaptation an engineering problem, evolutionary computation has shed light on how complex systems can be adaptive. Second, there is the rapid growth of distributed and network-mediated computing (including the Internet), which has led computer science into deeper analyses of just what it takes to make systems of many agents work together and grow.

From social design come insights into people and their activities in political, economic, and social systems. Entire disciplines—such as political science, economics, sociology, psychology, and history—have been devoted to understanding human beings and the settings they build and live in. Among the approaches that have concentrated on social design are organization theory and game theory. Organization theory provides insights into how institutional structure matters. Game theory provides insights into how people can choose strategies to maximize their payoffs in the presence of other people who are doing the same.

While the foundations of this work come from evolutionary biology, computer science, and social design, our analysis differs from all three of these in important ways.

Unlike evolutionary biology, we are primarily interested

- in the shaping of evolutionary processes rather than just observation and explanation,
- in intelligent individuals with language and culture, rather than plants and animals that rely primarily on their genetic heritage, and
- in different measures of success rather than taking the ability to have offspring as the sole measure of success.

Unlike computer science, we are primarily interested

- in systems composed of people or organizations rather than pieces of software,
- in systems with long and rich histories rather than systems that have little or no history, and
- in systems in which the costs of trials needed for adaptation are measured in terms of efforts and even lives of people rather than in cycles of computer time.

Unlike some approaches to social design, we are primarily interested

- in problems in which the preferences and even the identities of the participants can evolve over time, rather than situations in which the players and their preferences are fixed, as they are in game theory, and
- in problems in which decentralization is both promising and problematic, rather than situations in which decentralization is seen as practically a panacea, as in some forms of neoclassical economics.

In our analysis there are three key processes in a Complex Adaptive System. These key processes provide the basis of our three central chapters: Variation, Interaction, and Selection. We see variation, interaction, and selection as interlocking sets of concepts that can generate productive actions in a world that cannot be fully understood. We show how the very complexity that makes the world hard to understand provides opportunities and resources for improvement over time.

We are often asked how "complexity" differs from "chaos." The simple answer is that chaos deals with situations such as turbulence (Gleick, 1987) that rapidly become highly disordered and unmanageable. On the other hand, complexity deals with systems composed of many interacting agents. While complex systems may be hard to predict, they may also have a good deal of structure and permit improvement by thoughtful intervention.

Our approach is not just an extended "evolutionary metaphor," nor is it part of Social Darwinism (Hofstadter, 1955) or sociobiology (Wilson, 1975). We view processes of biological change as wonderful examples in the larger set of Complex Adaptive Systems. However, they have special kinds of agents, particular sorts of strategies, distinctive patterns of interaction, and their own special processes of selection. The patterns one sees in biology are not always found in other Complex Adaptive Systems. Copying a strategy for stock trading (such as a computer algorithm) involves only digital information and so is nearly costless compared with producing a new organism that contains a copied gene. Evaluating a business strategy (say, the introduction of a new product) can be enormously expensive compared with making a random variation of a fruit fly. Variation, interaction, and selection are at work in a population of business strategies, but the detailed mechanisms are often distinctly unbiological. To harness complexity effectively, many kinds of Complex Adaptive Systems must be considered.

We have paid special attention to the role of information in Complex Adaptive Systems. The continuing fall in the costs of copy-

ing and recombining information often results in the very rapid
spread of strategies. An increasing penetration of information technology into social processes will therefore change those processes
fundamentally.

We have emphasized the contextual forces determining interaction patterns. This too takes our work away from the traditional approaches in economics as well as biology, where there is often scant
attention to the important consequences of patterns of interaction.

These additional aspects of our framework make it richer, able to
incorporate realistic aspects of the world, though it still leaves out
many factors, as good frameworks must do. The richer framework is
not able to make detailed predictions, of course. There are too many
things that might happen. We are willing to bear the lack of detailed
prediction because we are interested in situations in which accurate
prediction has always been difficult. In return for accepting complexity, we have a more systematic approach to harnessing it. This
intellectual tactic reminds us of a guiding principle of the martial art
of judo: "Throw with your opponent's own weight."

Our emphasis on *harnessing* complexity will, we hope, prove to
be an important contribution. We chose "harness" for our title to
convey a perspective that is not explanatory but active—seeking to
improve but without being able fully to control. This orientation determined many features of the book but most basically the focus on
aspects of Complex Adaptive Systems that suggest guiding rules of
thumb and leverage points of intervention.

The Complex Adaptive System approach is a way of looking at
the world. It provides a set of concepts, a set of questions, and a set
of design issues. By itself, it is not a falsifiable theory. Such a theory
would have to specify the operational meaning of the key concepts
and mechanisms in a particular domain. For example, to apply the
Complex Adaptive Systems approach to economic markets, one
would have to specify who the economic actors are, what they can
see and do, how they generate variety in their behavior, how they interact with each other, and how the actors and their strategies are selected for retention, amplification, or extinction.

This book provides our personal view of how complexity research can be made relevant to problems of social design. We have benefited greatly from the many researchers working on complexity. Our task here is not to provide a textbook that surveys this dynamic field. Instead, we present our view of what complexity research offers to those who want to improve the world as well as marvel at it.

We are indebted to James March and John Holland for laying the foundations on which this book builds, and for providing personal as well as intellectual leadership. In the 1970s, James March began to write articles about a provocative topic: "the technology of foolishness" (March, 1976). He forced into the open an issue that remains hidden in a more conventional view of choice or decision making in social systems: the hard reality that the world in which we must act is often beyond our understanding. He began to draw out the implications of this fact when others were mostly in denial. It implies that each action we take is partly an instrumental step and partly a learning experience.

From John Holland we learned how adaptation can be regarded as an engineering problem. His inventions, starting with the genetic algorithm (Holland, 1975), provided a systematic way to design and study Complex Adaptive Systems with computer simulations.

For providing a wonderful network of connections to fellow workers in complexity theory, we are indebted to the Santa Fe Institute and the University of Michigan's Center for the Study of Complex Systems.

The immediate impetus to write this book was provided by a report commissioned from us on national information policy by the Highlands Forum. The Highlands Forum is a group of people from industry, government, and academia that deals with information issues and is sponsored by the Department of Defense. We decided to write our report from the point of view of what the Complex Adaptive Systems approach can say about information policy. In writing that report, we developed our own vision for how complexity can be harnessed in information policy. We then saw that this vision could be applied to many different areas of social design.

An indirect source of inspiration for this book was provided by a project sponsored by the Intel Corporation through a grant to the two of us and Rick Riolo. This project uses large-scale computer simulation experiments to study self-organizing social structures. The work we did on this project, along with the modeling work that the two of us have been doing for several decades, helped us better understand how complex systems can be built and analyzed.

For valuable advice (not always followed) we would like to thank David Axelrod, Arthur Burks, Corinne Cohen, Rachel Cohen, Paula Duffy, Edmund Durfee, George Furnas, John Holland, Christopher Lee, Ann Lin, James March, Jeffrey Mackie-Mason, Melanie Mitchell, Stephen Morrow, Scott Page, Paul Resnick, Rick Riolo, Douglas Ross, Raphael Sagalyn, Amy Saldinger, Carl Simon, Robert Wallace, Michael Wellman, and Marina von Neumann Whitman. For editorial help we thank Maria Bonn, Loretta Denner, and Donna George. For financial assistance, Robert Axelrod thanks the University of Michigan LS&A Faculty Enrichment Fund. Michael Cohen thanks the Ameritech Foundation.

For continual inspiration over more than fifteen years, we are indebted to the BACH research group at the University of Michigan. The BACH group is named after its original members: Arthur Burks, Robert Axelrod, Michael Cohen, and John Holland. The group also came to include long-term members William Hamilton, Rick Riolo, and Carl Simon and shorter-term members Stephanie Forrest, Douglas Hofstadter, Melanie Mitchell, Michael Savageau, and Reiko Tanese. Like any good complex system, it has become more than the sum of its parts. We are delighted to dedicate this book to the BACH group.

HARNESSING
COMPLEXITY

I

Introduction

Whether or not we are aware of it, we all intervene in complex systems. We design new objects or new strategies for action. They may be as simple as a paper form for recording the hours spent on a team project, or as sophisticated as an "artificial agent," a free-roaming computer program empowered to buy and sell goods with real money. We also devise policies. They may be as simple as changing the rules for making very small bank loans, or as sophisticated as forging an international human rights treaty. Whether simple or sophisticated, such actions change the world and, as we will shortly see, can lead to consequences that may be hard to imagine in advance.

There are, of course, many situations that are not complex, where we know what actions are possible and what consequences they will likely produce. When we do, we can choose the action that seems most promising, perhaps recognizing some risk that an unwanted outcome could occur. We order clothing from a catalog. We expect to like what we receive, though it could happen that we do not.

In many design and policy settings we face a very different situation. We sense that a choice may have large consequences, but we are unsure what these may be. For example, in recent years people have asked: "What might happen if we made small groups responsible for loans instead of individuals?" and "What might happen if we made a simple change to twenty-year-old Internet protocols that would allow hypertext links among files called Web pages?"

This book is about designing organizations and strategies in complex settings, where the full consequences of actions may be hard—even impossible—to predict. It draws together ideas from scientific research done in recent decades on many kinds of complex systems. The book does not provide a magic lens that will suddenly make the cloudy future clear. The complexity of the world is real. We do not know how to make it disappear. What the book does provide is a framework, a way of thinking through a complex setting that takes advantage of complexity to generate new questions and new possibilities for action. It suggests ways of "harnessing complexity."

In this first chapter, a set of examples briefly introduces many of the main ideas of our framework. We do not linger over important details that are considered later, but we set out the central ideas that have guided our development of the framework. They are:

- the difficulty of prediction in complex settings,
- how related themes of complexity have arisen in the physical, biological, and social sciences,
- how concepts from complexity studies can be useful in settings whose consequences are hard to predict, and
- the main mechanisms and design principles that we identify from complex systems research.

Our framework synthesizes these mechanisms and principles into a coherent approach to complex systems. It provides a device for channeling the complexity of a social system into desirable change, just as a harness focuses the energy of a horse into the useful motion of a wagon or a plow. The introductory chapter includes a discussion of the current "Information Revolution." This profound transformation is historically and intellectually entwined with complex systems issues and is therefore a natural source of examples throughout the book. The chapter closes by considering how complexity provides a profoundly different way of thinking about social systems compared with the more familiar perspective inherited from the Industrial Revolution. We develop the framework in the three chapters that follow our introduction: Variation, Interaction, and Selection. In them we study important mechanisms that can be used to influence the amount of variety in a system so as to affect the balance between exploration and exploitation, to alter the structure of interactions within the system, and to adjust the way success is measured and amplified.

Introduction to the Framework

The framework that we elaborate in the subsequent chapters views complex systems in terms of populations of agents. To get a feel for how it works, consider these situations:

- A member of a work team wants to elicit cooperation from co-workers to get a job done;
- An impoverished woman in Bangladesh needs to borrow a little money from someone to start a stall in the local market;
- A computer program scans the Internet seeking helpful resources;
- The United States wants to foster among nations goals it cannot impose by force, such as human rights, openness to international trade, peaceful resolution of conflict, democratic governance, and market economies.

As varied as these settings are, they can be approached in a common way. We see all of them as complex systems in which interventions could induce large changes. We can represent them all within a common vocabulary, drawn from the key terms of our framework.

The first concept is that of an **agent.** (As we introduce each of the major concepts in the book, we distinguish it with boldface type.) An agent has the ability to interact with its environment, including other agents. An agent can respond to what happens around it and can do things more or less purposefully. Most commonly, we think of an agent as a person, such as the team member in a company or the person seeking a loan. Considering this broad definition, we can see that a person is not the only kind of agent. A family, a business, or an entire country can also be an agent. Even a computer program interacting with other programs can be regarded as an agent.

When we talk about agents we will usually expect them to have a number of properties. These include location—where the agent operates; capabilities—how the agent can affect the world; and memory—what impressions the agent can carry forward from its past.

The second key concept is **strategy,** the way an agent responds to its surroundings and pursues its goals. An employee might help a co-worker in the hope that the co-worker will reciprocate. Someone needing a small loan might ask friends to help out. A nation seeking to promote favorable norms might try to lead by example. A computer program seeking useful resources might buy information from other programs and keep track of which ones provided resources that were actually worth what was paid. These are all strategies. Our usage includes deliberate choice, in the sense of the term "business strategy," but it also includes patterns of response that pursue goals with little or no deliberation.

A central interest of ours is how strategies change over time. One source of change is the agent's experience of how well the strategy is doing. An employee, finding that co-workers are not contributing to a joint project, might decide not to contribute either. Typically, human agents have some awareness of their own strategies, and they may be able to observe something about how well they are doing according to some **measure of success.** Often they can observe the actions or successes of the agents with whom they interact. This may trigger a dissatisfied agent to try a new strategy based on trial and error, or to imitate the strategy of another agent.

Changes in strategies can also come about through changes in the population of agents. For example, experienced workers may train new workers, or practices at one company may be imitated at another. Such processes of reproducing, or **copying,** play an important role by changing the mix of strategies or agents in a population.

The idea of a **population** of agents is our third major concept. Indeed, the idea is so central that we sometimes refer to our framework as the "population approach to Complex Adaptive Systems." If you are seeking to harness complexity, populations are important in three ways: as a source of possibilities to learn from, as recipients for a newfound improvement, and as a part of your environment. For example, if you are a business manager, you can learn from the population of managers who face similar problems, you can spread what is learned to a population of co-workers, and you can see your company as one part of a population of businesses and consumers that you adapt to even while they are adapting to you. You can think about populations of strategies as well as populations of agents. For example, if you try different ways of raising funds for your nonprofit organization and observe others doing fund-raising for theirs, you can learn from the resulting population of fund-raising strategies. One of the key benefits of the Complex Adaptive Systems approach is that it helps you see yourself in the context of a population of agents, and helps you see your actions in the context of a population of strategies.

Many of the key questions generated by our framework center on the way strategies or agents of a particular **type** become more (or less) common in a population. For example, "aggressive" and "low-key" might be types of sales strategies that a particular firm distinguishes. Another firm might distinguish "recurring" from "onetime." Teachers might define the population of children at their school (agents, in our terms) as falling into types by grade levels. For other purposes, genders might be the relevant types.

Our rough criterion for the boundaries of a population will be that two agents are in the same population if one agent could employ a strategy used by another. So, for example, a villager might try an approach to borrowing money that had been effective for a family member or friend.

This simple example of villagers reveals two important features of populations. First, strategies spread (and sometimes change) by moving among members of a population. A borrowing strategy might spread by word of mouth through family or friendship networks. It might also change in some significant way as it is repeatedly retold. Change processes such as this create **variation** among strategies. Second, populations have structure—**interaction patterns** that determine which pairs of agents are likely to interact and which pairs unlikely. The borrowing strategy moves among friends or relatives.

Real situations often include more than just a single population of agents, of course. For one thing, there may be several different kinds of agents. There are not only sellers in the village, but also buyers. There are not only nations in the international system, but nongovernmental organizations. Moreover, many settings include important entities that are not agents. Books, vehicles, weapons, and medicines play significant roles even if they lack some qualities of agents. We will be especially interested in **artifacts,** objects that are used by agents. Like agents, they can have important properties, such as location or capabilities. A toy may respond to a child who winds its spring. Artifacts may have "affordances," features that evoke certain behavior from agents, like the beautiful handle of a pitcher that invites the grasping hand. However, artifacts usually do not have purposes of their own, or powers of reproduction.

When we want to talk about a real situation, we will generally pack all of these elements up into the term **system.** We will use the word to indicate one or more populations of agents (for example, employees and customers of the company), all the strategies of all the agents (working together to produce and sell products, and buying and using products), along with the relevant artifacts and environmental factors (manufactured products, production tools, sales brochures, and store opening hours).

Whenever we are interested in designing something new (such as a product or sales strategy), or when we are contemplating a possible change in a policy (such as new store opening hours), we are considering interventions in a system. But what might make a system we are interested in complex? This is a question to which we will be returning, but let us begin by saying that a system is **complex** when there are strong interactions among its elements, so that current events heavily influence the probabilities of many kinds of later events.

A major way in which complex systems change is through change in the agents and their strategies. We saw earlier that there are many processes of strategy change. We will be interpreting them as many different forms of **selection.**

Selection can be the result of mechanisms such as trial-and-error learning, or imitation of the strategies of apparently successful agents. Selection can also result from population changes like birth and death, hiring and firing, immigration and emigration, or start-up and bankruptcy. Selection need not always be beneficial. Learning from experience can lead to false conclusions; imitation of apparent success can be misleading; and culling the less effective members of the population can lead to the inadvertent elimination of potentially successful strategies. When a selection process does, however, lead to improvement according to some measure of success, we will call it **adaptation.** Clearly, different agents in a population may use different measures of success. So changes that are adaptations for some may not be for others.

When a system contains agents or populations that *seek* to adapt, we will use the term **Complex Adaptive System.** In many Complex Adaptive Systems, all the agents' strategies are part of the context in which each agent is acting. This makes it hard for an agent to predict the consequences of its actions and therefore to choose the best course of action. Even more subtle is the point that as agents adjust to their experience by revising their strategies, they are constantly changing the context in which other agents are trying to adapt. For example, workers in a company might be learning better ways to produce a product. Each change of strategy by a worker alters the context in which the next change will be tried and evaluated. When *multiple populations* of agents are adapting to each other, the result is a **coevolutionary process.** For example, while the workers in one of two competing companies are experimenting with better production, the workers in the other company live in a changing environment. And their efforts to adapt may change the context of improvement efforts in the first company. This can lead to perpetual novelty for both sides. The system may never settle down.

The impoverished woman seeking a loan is also in a Complex Adaptive System consisting of many others: potential borrowers and creditors, merchants and consumers. Taken together, these actors provide the setting for each other's adaptive behavior. Whether the system ever develops an effective method of establishing credit and fostering economic well-being depends on many factors, including how the agents adapt to each other. The United States is also in a Complex Adaptive System. Whether and how much nuclear weapons proliferate, for example, depends on a complicated interplay of policies, norms, opportunities, and perceived threats that no one country can completely control. A computer program may live in a world of other programs. What makes it successful in achieving the needs of its user depends in part on actions of other programs it meets and on how they adapt to each other.

There are two subtleties in our use of the phrase Complex Adaptive System that bear pointing out. First, we use the phrase when the agents *may* be adapting. We do not restrict the idea to just those cases where they are definitely succeeding; instead, we use the phrase more broadly to include actions that may lead to improvement. Second, our use of the term says only that the *parts* are adapting, not necessarily the *whole*. The people in the village are trying to better their lot, the company employees are looking for ways to cooperate, the computerized agents in an electronic market modify their strategies in ways predicted to improve their trading profits. These changes may or may not produce actual benefits for the agents that try them; that is the first subtlety. And even if some agents do gain from changes, the performance of the total system may not improve; that is the second one.

An important reason we do not require that either the agents or the system be succeeding is that we want to help foster future adaptation. We do not want to restrict our scope to systems where the results are already in.

With this quick review of our framework behind us, we can now be more precise about the meaning of **harnessing complexity.** The phrase means deliberately changing the structure of a system in order to increase some measure of performance, and to do so by exploiting an understanding that the system itself is complex. Putting it more simply, the idea is to use our knowledge of complexity to do better. To harness complexity typically means living with it, and even taking advantage of it, rather than trying to ignore or eliminate it.

Let's return to the four examples introduced earlier to show how complexity can be harnessed in a variety of ways.

A member of a work team seeking to promote contributions of time and effort to a joint project might set up a way for each worker to know what the others contribute. This would allow recognition of individual contributions. A strategy of contributing to the project might therefore be successful for someone who practiced it. Others might then copy this type of strategy. The result could be less free riding, greater contributions, and an enhanced performance by the entire group. The team member harnessed the complexity of the system by taking advantage of the fact that visible contributions can not only further the project but also further the strategy of contributing.

A Bangladeshi economist named Muhammad Yunas had a wonderful idea to help poor people obtain small loans (Bornstein, 1996; Shenitz, 1997; Yunas, 1999). Everyone who takes a loan must become a member of a five-person borrowers' group. The groups share responsibility for loan repayments or defaults. The five members of a borrowers' group agree to take joint responsibility for a loan to one of their number knowing that only if the loan is repaid can another member of the group get a loan.

The system is called Grameen banking, after the Bangla word for "village." The idea is so effective that 97 percent of loans are repaid, which is comparable to Chase Manhattan's rate. Today there are over a thousand branch offices, serving more than two million clients in Bangladesh, and the idea is being imitated in many countries, including underdeveloped regions of the United States.

In our terms, Grameen banking harnesses complexity by using existing social networks in a new way. When five potential borrowers get together, they engage in a new kind of interaction involving getting and repaying loans. The success of Grameen banking is built on the knowledge and interdependence that the members of the borrowers' group already have with each other. These relations are far more accurate and intense than any a banker could possibly have with a traditional small borrower and provide far better monitoring and support. Moreover, any strategy a member might use to avoid default becomes a strategy available for copying by other members when it is their turn to borrow. Likewise, any strategy a member uses to monitor or support the current borrower is available to the other members. The very complexity of existing village networks is harnessed by the Grameen banking system for the purposes of increasing available credit and thereby promoting small business.

Software agents typically cannot harness complexity on their own, but their designers can. A powerful technique that harnesses complexity is called the genetic algorithm (Holland, 1975; Mitchell, 1996). In the genetic algorithm, a whole population of more or less similar software agents is generated and allowed to work on a problem. Each gets a score for its work based on some measure of success. The relatively effective ones are allowed to reproduce themselves. The less effective are discarded. This is a form of selection.

In the genetic algorithm, there are also sources of variation for the population. Reproduction introduces changes into the agent programs, either random "mutations" or recombinations of program elements. These changes alter the population of software agents, and over time the agent programs in the population become better able to solve the problem at hand. Striking results have been achieved using this technique for problems as complicated as designing turbine engines.

The United States can exploit the complexity of the international system in many ways, but one of them is to set an example in its own behavior that, if emulated by other agents, would improve the international system. Precisely because the international system is so complex, it is hard for any country (or other transnational actor) to determine what is in its own best interests. So a reasonable tactic for many nations is to copy the observed behavior of large, apparently successful actors such as the United States.

In the coming chapters on variation, interaction, and selection, we examine more closely the elements of our framework briefly set out here. In the remainder of this introduction we elaborate the themes we raised at the outset: the difficulty of predicting in complex settings, and how complexity research may help us to intervene effectively in complex systems with adaptive agents.

The Difficulty of Prediction

Although we all do our best to foresee important consequences, there is widespread acknowledgment that this is extraordinarily hard in times of dramatic change. The Information Revolution provides excellent examples, for deep reasons we will examine. Some of the most famous stories of mistaken foresight center on managers and board members at companies like IBM and Intel who were unable to grasp the world-changing potential of their own products. IBM leaders once thought a handful of computers would suffice for the entire world. The Intel board of directors discouraged the first proposals to develop a microprocessor. The National Science Foundation has remarked that its panel of distinguished information technology scientists and engineers is consistent in its unwillingness to predict the future (*New York Times,* 1997). Efforts of the Justice Department to redress the consequences of Microsoft Corporation's monopoly are hampered by the inability of experts to say what operating systems might become. As Andrew Pollack said, "The gears of the digital revolution [are] turning faster than the wheels of justice" (Pollack, 1998). Some industry leaders were frank enough to say—two years after the deluge—that they saw the first effective Web browser, Mosaic, as an inconsequential toy (Norman, 1997). As we write, that experience of the unanticipated World Wide Web explosion is fresh in our memories. In the Information Revolution, there are clearly strong limits on our ability to foresee what is to come.

A wary attitude toward prediction is probably healthy, but it presents a severe roadblock to the normal processes of designing new artifacts or strategies, or refining and implementing policies. The standard procedure of design and policy making is to develop expectations (predictions) of how the future will unfold, and to define actions we could take that would lead to more desirable predicted futures. This stance can be stretched to accommodate some uncertainty by bringing in specialized techniques like Bayesian inference to deal with probability distributions on possible futures. But the usual approaches to designing an intervention grind to a halt if we acknowledge that we do not know what might happen as a consequence of our actions.

When experts are asked to forecast the future and its requirements in complex settings, their customary response is to acknowledge the difficulty of prediction and then do the best they can with their particular expertise. This strategy of making a "best guess" is entirely sensible. In many situations, any one of several actions is better than no action.

Retrospection also shows that in hard-to-predict moments often *someone* did identify what would happen and had a good sense of what could have been done. For example, in the domain of information technology there was an early vision of what became the World Wide Web (Bush, 1945). But a careful observer of such moments also sees that there were usually many conflicting expert predictions in play, even for the effects of a single factor. Before the fact, there was no reliable way to discern which prediction would turn out to be right. Even though something may be better than nothing, and even though someone among the contending experts may be right, it can still be very disquieting to act on the "fiction" that we have a reasonable prediction of the consequences of a particular event in a complex system.

An alternate line of response to the difficulty of prediction is offered by various forms of scenario generation (Wilkinson, 1995). This is the exploration of what are thought to be major driving forces of the situation, looking for policies that are robust even if there may be changes in currently dominant factors. By encouraging structured thought about the future, scenario generation tries to secure the benefits of preparation: being ready with some plausible response as the unexpected unfolds. The approach still requires an ability to identify correctly the principal driving forces in the system, and how they will affect the outcomes of interest. The approach is hobbled if we cannot answer these questions. For example, scenarios for future political development in poor countries depend on whether current social structures will hold together or fall apart. Social structures are likely to be affected by the clear trends of increasing deployment of information technology in these countries. Scenario development may be impossible if we cannot say how a process such as fragmentation of social structure will be affected by technologies such as rural cellular telephones and satellite television because the driving forces remain obscure.

We believe that the difficulty of prediction in complex systems does not make the situation hopeless, although it does require a large shift in the conceptual tactics. The framework we develop here can complement and strengthen conventional and scenario-building approaches to changing the future of complex systems. To see how the framework can help, we need to consider the role of complexity in prediction difficulty, and the ideas from complexity research that can be applied in response.

What makes prediction especially difficult in these settings is that the forces shaping the future do not add up in a simple, systemwide manner. Instead, their effects include nonlinear interactions among the components of the system. The conjunction of a few small events can produce a big effect if their impacts multiply rather than add. The overall effect of events can be unforeseeable if their consequences diffuse unevenly via the interaction patterns within the system. In such worlds, current events can dramatically change the probabilities of many future events. A collection of complex systems is therefore a kind of dynamical zoo, a "wonder cabinet" of processes that change (or resist change) in patterns wildly unlike the smoothly additive changes of their simpler cousins. The complex systems world is a world of avalanches, of "founder effects" (where small variations in an initial population can make large differences in later outcomes), of self-restoring patterns (in which there can be large disturbances that do not ultimately matter), of apparently stable regimes that suddenly collapse. It is a world of punctuated equilibria (where periods of rapid change can alternate with periods of little or no change), and butterfly effects (where a small change in one place can cause large effects in a distant place). It is a world where change can keep recurring in a fixed pattern, where rapid and irreversible change can occur when a certain threshold of effect is reached, and where great variety can exist at a large scale, even though small patches have very little variety. These are not completely disorderly worlds, so turbulent that useful lessons can never be learned. They have structure, and beneficial adaptation can sometimes occur. But prediction and choice of the conventional kind are not very reliable.

It is worth noting that the difficulty of predicting the detailed behavior of these systems does not come from their having large numbers of components. Often science can skillfully exploit large numbers to get better predictions than would be possible for smaller systems—think of the gigantic number of colliding molecules in a perfect gas, where pressure, temperature, and volume conform to Boyle's Law. Conversely, there are some notable complex system models, such as Conway's "Game of Life," where very complex behaviors arise from the interactions among small numbers of extremely simple elements (Gardner, 1970).

For us, "complexity" does not simply denote "many moving parts." Instead, complexity indicates that the system consists of parts which interact in ways that heavily influence the probabilities of later events. Complexity often results in features, called **emergent properties,** which are properties of the system that the separate parts do not have. For example, no single neuron has consciousness, but the human brain does have consciousness as an emergent property. Likewise, a uniform price can emerge in an efficient market of many buyers and sellers. Research in recent years has begun to develop a literature on emergent properties and other characteristics of complex systems as a class (see, for example, Belew and Mitchell, 1996). To distinguish systems that do have a lot of "moving parts" but may not be complex, we will use the term **complicated.**

Complexity Research

At this time, there is little convergence among theorists who have begun to study complex systems as a class. It is not a field in which a crisp and unified theory has already been developed, nor is one expected in the next few years. For example, there is no agreement yet on the best way to measure the amount of complexity in a given system (Johnson, 1997). In many ways we are sympathetic to a proposal of Murray Gell-Mann's (1995), which captures the sense that a system should be called complex when it is hard to predict not because it is random but rather because the regularities it does have cannot be briefly described. This distinguishes complexity from randomness, and it aligns with our focus on difficulty of prediction. But there are a number of other careful definitions of complexity that have other desirable properties (for example, Bak, 1996; Cowan, et al., 1994). If there will be a consensus on a precise definition of complexity, it lies well in the future.

The many definitions reflect the recent history of complexity research, which involves a remarkable diversity of domains. Each such area has specialized characteristics that distinguish it to some degree from the others. At the same time, there are recurring deep themes about agents, strategies, interactions, and copying that link these diverse fields.

The fields within which complexity research has developed include:

- Condensed matter physics, with its concern with nonlinear interactions among the spins of many particles and the retention of patterns that reduce "frustration" between the elements of a system. (This is the class of problems that led Nobel laureates Murray Gell-Mann and Philip Anderson to help found the Santa Fe Institute, which has since become the leading center for the study of complexity in all its manifestations.)
- Evolutionary biology, with its concern for populations with "gene pools" of strategies that evolve through selective reproduction with variation.
- Evolutionary computation, the branch of computer science that is inspired by evolutionary biology to develop techniques that can discover good strategies for hard problems by means of populations that reproduce with variation and selection (Holland, 1992).

- Social science modeling of heterogeneous populations of people who interact with each other and keep or change their strategies depending on how well they are doing (Schelling, 1978, especially pp. 147–57).
- Cellular automata, populations of very simple, locally connected computing elements, including the extremely clever example invented by John Horton Conway, the "Game of Life" (Gardner, 1970; Poundstone, 1985).
- Artificial life, the study of many different systems, usually implemented as computer-simulated agents, that exhibit lifelike properties such as self-reproduction (Langton, 1988).
- Mathematical theories to formalize the measurement of the complexity of a system (Lloyd, 1990; Gell-Mann, 1995).

Despite the diversity of these areas, we can identify recurring themes in the work of complexity researchers. A number of these themes can be distilled and brought to bear on the problem of analyzing interventions in a world that is hard to predict because it is complex.

In contrast, a world that is hard to predict merely because it is complicated can be attacked in quite a different way. For example, nearly additive contributions of factors mean that independent studies of the important factors can later be merged at acceptable cost. The Human Genome Project is a large bet that much can be understood via such a "divide and conquer" strategy. Nevertheless, in many cases the interactions of the parts of the system are critical, and complexity reigns.

Our framework can be developed to give a unified view of the work on complex systems. As we look across many lines of this research, we see in most of the studies collections of elements—what we have called populations of agents. Usually those elements subdivide into some types (for example, buyers and sellers, inhibiting molecules and potentiating ones, Kosovars and Serbs). Each of the elements is connected to some, but usually not all, of the others. The connections are through relations, and there is tremendous variety across fields in what those relations are and how they work (for example, magnetic attraction, organizational authority, electrical stimulation, sexual affinity, chemical inhibition, geographical proximity, or ethnic hostility). Each element in one of these complex systems has patterns of action that affect those connected to it.

The research very often centers on the emergent global dynamics of the whole system. It asks questions like: How (or when) does a system of locally trading agents develop prices that will cause marketwide inventories to clear? How does a brain made of interconnected neurons learn? How does a pile of sand generate its characteristic mix of large and small avalanches? How does news about a vacant job successfully make its way from an employer to potential employees in distant towns? How does a gene pool remix itself over time to create and retain genotypes that may be fit for a changing environment? How do we nurture a network of trust that permits informal credit mechanisms to foster trade efficiencies?

It is usual in this approach to view the global properties of the system as emerging from the actions of its parts, rather than seeing the actions of the parts as being imposed from a dominant central source (Holland, 1998). This is not a denial that there are times when systems have effective central authorities or dominant influences. But the project of complexity theories in such cases is to understand how those dominant influences come about, what sustains (or undermines) them, and how local action responds in the face of global constraints. An excellent example is the work of Padgett and Ansell (1993) on the emergence of a new form of state in medieval Florence as Medici power built up out of tensions within the marital, residential, and commercial networks of the city.

Finally, many complex systems—but by no means all—are "adaptive." As we said earlier, in systems we call adaptive the strategies used by agents or a population change over time as the agents or population works for improved performance. When we use the phrase Complex Adaptive System, we leave open the question of whether the agents or population actually achieves improved performance. If we are designing interventions, improvement on some measure is what we want to promote. For a system to exhibit adaptation that enhances survival (or another measure of success), it must increase the likelihood of effective strategies and reduce the likelihood of ineffective strategies. We call such a process **attribution of credit** if an agent uses a performance criterion to increase the frequency of successful strategies or decrease the frequency of unsuccessful ones.

Complexity research has received considerable attention recently. In some measure, this is because advances in computation have enabled progress on a number of problems that had long been too difficult for conventional mathematical tools. But it is important to recall that the fundamental orientation of complexity research is actually rooted in long traditions. Adam Smith's hidden hand, the "blind watchmaking" of Darwinian evolution, the cell-assembly neuropsychology of Donald Hebb, and the self-reproducing automata of John von Neumann were earlier intellectual developments that blazed the same trail by uncovering system-level properties produced by the structured interaction of simpler components.

Perhaps there are powerful results just over the horizon, but as we see it, complexity research does not make detailed predictions. Rather, it is a framework that suggests new kinds of questions and possible actions. We would compare the results taking shape to the artificial selection principles of animal husbandry (a field that much interested the youthful Darwin). Analyzing complex systems within the framework does not assure the ability to produce specific outcomes but can foster an increase in the value of populations over time—whether the populations are of livestock, of technical innovations, or of new strategies for business competition.

In the language of our framework, a **designer** introduces new artifacts or strategies into the world. A new machine on a factory floor or a new approach to conducting a budget review may be interventions in complex systems whose full consequences cannot be contemplated in advance. An orbiting telescope and a legal appeal on constitutional grounds almost certainly will have consequences that are hard to predict. A designer may even introduce new agents into the world. For example, an executive might create a new division in an organization, or a legislature might set up a new governmental bureau. **Policy makers** deliberately alter the consequences of available strategies when they increase rewards for some outcomes or make some patterns of action illegal. We use the phrase "design and policy making" to indicate the full spectrum of actions that we may find ourselves considering.

We may take the perspective of someone within a system—for example, as one of many people at a committee table. Alternatively, we may contemplate the system from the outside, as an architect or a legislator might do. In either case, we all find ourselves designing or making policy in complex settings. When we do, it can be very valuable to extend the questions we conventionally ask about likely consequences and scenarios. We can go on to ask what populations of agents and strategies are involved, and what interventions might create new combinations or destroy old ones? These kinds of questions help us harness complexity.

The Design of Organizations and Strategies

Why might this perspective and its associated vocabulary be useful for deliberations over action in settings whose consequences are hard to predict? Complexity research deals with systems that are hard to control, and much of it has gone on in fields that seem far from policy or design concerns. (Do the neurons fire in synchronized waves or incoherently? Are the magnetic poles of particles oriented like those of their near neighbors or are they disarrayed? Do most of the animals in a population continue to exhibit a certain useless trait, or has this type vanished over time?)

However, many of the same dynamics are involved in social issues. (Do similar transactions across the economy take place at one
price or many? Does animal husbandry improve the agricultural
value of a livestock population? Do citizens remain loyal to a single
large state, or transfer their loyalties to smaller political units? Does
an infection—or an Internet rumor—become endemic in certain subpopulations?)

Social systems exhibit dynamic patterns analogous to physical,
biological, and computational systems. This is perhaps the fundamental reason we pursue complexity research. Many social interventions
are directed toward controlling the interaction among types of agents.
For example, segregation (and integration) of races; visa and immigration rules; entry qualifications to religious and social organizations; "cultural revolutions" and "peace corps" that send the highly
educated to less developed areas; political redistricting; zoning restriction of commercial, industrial, and residential activities; film,
television, and Internet ratings to facilitate matching of audiences and
contents; and foster-care systems that place children with adults dif-
fering from their parents in race or class. Many other policies have
important or interesting side effects that are related to interactions of
types. For example, imprisonment that mixes experienced criminals
with rebellious adolescents; public transit patterns that separate urban
center residents from suburban jobs; armies of occupation that result
in children of intermarriage; computer networks for defense and science that increase communication between parents and college-
distant children and facilitate finding of long-lost friends.

Complexity research gives us a grounded basis for inquiring
where the "leverage points" and significant trade-offs of a complex
system may lie. It also suggests what kinds of situations may be resistant to policy intervention, and when small interventions may be
likely to have large effects. For guidance in designing actions, such
insights into the right questions can be very valuable. They can be
valuable even if the theories are too multiple and too preliminary to
support any claim that a theory of complexity implies any sharply

etched expectation about a future scenario and how a particular action will guarantee it.

We are hardly the first to sense the promise of complexity research for guiding action. Books, consulting services, conferences, and journals have been developed to respond to the intuition of many managers and professionals that there is a deep resonance between concepts in complexity research and the problems of designing effective interventions. We note books such as Mitchell (1996) and Kelly (1994). Regular conferences for managers have been organized in recent years by the New England Complex Systems Institute, among many others; examples of journals include *Complexity* and *Emergence: A Journal of Complexity Issues in Organizations and Management.*

Our contribution lies in our attempt to move the work beyond metaphorical affinities and to distill an explicit method that can be applied in practice. At the same time, we want to avoid simplistic lists of four (or nine, or twelve) principles that readers should always follow. As we see it, unqualified nostrums could hardly be less in keeping with scientific work on complexity.

Our aim is to build on the resemblance between change processes seen in the systematic study of Complex Adaptive Systems and potential sources of change in social systems. It is our argument that principles derived from working with complexity problems shed valuable light on the issues confronting policy makers and designers.

We develop this argument by characterizing a set of change mechanisms in complex systems. While we cannot hope to provide any sort of exhaustive catalog, we can show that a wide array of examples fall into a few useful clusters. Most of the mechanisms and related principles that have policy relevance center on three central and deeply connected questions:

1. What is the right balance between variety and uniformity?
2. What should interact with what, and when?

3. Which agents or strategies should be copied and which should be destroyed?

These are necessarily rather abstract issues. In the next three chapters, we develop them by expanding on each one in turn. In each case we provide examples from complex systems studies, along with examples from social systems with more intuitive connections to design and policy concerns. We discuss similarities between crossover mechanisms that recombine genetic materials and deliberate invention activities that recombine high-order concepts—such as motor and wagon to make the first "horseless carriage." We go on in each chapter to summarize observations about the recurring useful concepts that have developed as a result of complexity research in the area.

The Information Revolution

In the coming chapters, as in this introduction, we make heavy use of examples derived from the Information Revolution that is surging all around us. These developments are triggers for an enormous range of design and policy actions with which we all find ourselves engaged. That alone would make them useful material for our discussions. But there are deeper connections of the Information Revolution to a framework for harnessing complexity. The Information Revolution will provide excellent illustrations of many of our key concepts.

Our society is enmeshed in a major social transformation, driven in part by, and deriving much of its distinctive character from, the amazing advances in technologies of information. The rate of technical change in processing, storage, bandwidth, sensing, and effecting is dizzying. The technical changes in turn facilitate large shifts in most of our fundamental institutions: in nation-states, communications industries, churches, armies, factories, friendship networks, and more. The rate of social change is intoxicating, disorienting, and probably accelerating.

Throughout the book, we refer to the transitions underway as the "Information Revolution," although forces other than information technology are also deeply involved. Transportation, biotechnology, marketing, and a host of other technologies have expanded dramatically in the last half-century. Information technology has fueled these expansions and been shaped by them. While acknowledging these complications, we concentrate here on the *Information* Revolution, the aspect of our era that seems to us to have the most novel and transforming properties.

An Information Revolution seems to demand policy interventions at every level of social organization. What shall nation-states do about encryption or boundary-spanning financial crimes? What shall families do about unsavory materials their children can easily access? What shall armies do to prepare for attacks on "infostructure"? What shall charitable organizations and business firms do about the privacy of records kept on their clienteles? In all these cases and thousands more, deep questions are being asked about how interventions—designs and policies—can steer future developments in beneficial directions. In an era in which so many customary social, political, and economic arrangements seem up for grabs, what interventions will bring us to a future we would prefer? In all of these settings, people often ask how the likely consequences of actions can be foreseen.

Using our framework, we depart sharply from conventional efforts to foretell the future and draw policy implications for the unfolding Information Revolution. Instead, we offer a way to analyze the situations and intervention possibilities that flow unceasingly from this enormous change. Throughout the book, we offer purposeful questions where general answers cannot be known.

We can start by considering why at this historical and technological juncture we should *expect* the future events in an Information

Revolution to be especially difficult to discern. Our answer lies in the *complexity* of the social and technical processes whose rates of change are accelerated. Our view of the relevance of work on Complex Adaptive Systems can be grounded by examining two arguments. They relate the concept of information to complexity and to adaptation.

Complexity and Information

It has become widely accepted that a major source of prediction difficulty in the contemporary Information Revolution is the multiplicity of forces that are *interacting*. For example, the hard lesson has been learned that technologies are adopted not only as a function of cost, but also as a function of numbers of others adopting. A technology with a small market lead may become dominant even when it is not superior in quality, as in the stories of VHS versus Beta, and the QWERTY keyboard (David, 1985). Effectiveness of technology has been observed sometimes to depend on deployment of other technologies, such as Internet service provision depending on the installed base of telephones.

There have been striking cases of process surprise, such as a way of replacing carbon paper (xerography) that can upset the internal security of autocratic nations, as well as alter the conduct of basic office procedures. And cultural variables have been shown to set a controlling context for technical developments, as in rural areas of developing countries that may leapfrog wired communications to go directly to wireless, or when countries with nonalphabetic languages have sharply different approaches to word processing. Reaping the benefits of new technology has turned out often to require collateral resources, so that innovations imagined to favor equality could turn out to accelerate differences between social classes. We have learned that the absence in electronic mail of socially controlling status cues

can unleash embarrassing episodes of "flaming," in which participants write things they would never say to a recipient's face (Sproull
and Kiesler, 1992).

Such lessons have taught us all that virtually every important
force in collective life affects the way the Information Revolution
plays out. Scale economics, technological preconditions, national
developmental sequencing, social status, economic inequality, internal security postures, cultural context, and many more forces work
to *condition* the development of information technology impacts.
This is not unique to the contemporary episode in the growth of information technology. The earlier episode of the Information Revolution that began with movable type also had epochal consequences.
The historian William McNeill points out that the Chinese Empire,
the Islamic states, and the Christian West each gave its own distinctive shape to the movable type revolution in printing (McNeill,
1996). Roughly speaking, the Chinese used printing to reinforce
central authority, while Islam suppressed the technology. The Western case has been highly interesting to scholars because of its many
indirect effects including contributions to the promotion of religious
conflict and the rise of nation-states (Anderson, 1983; Eisenstein,
1983). These are just the kind of nonadditive contextual effects that
distinguish complex dynamic regimes.

If complexity is often rooted in patterns of interaction among
agents, then we might expect systems to exhibit increasingly complex
dynamics when changes occur that *intensify* interaction among their
elements. This, of course, is exactly what the Information Revolution
is doing: reducing the barriers to interaction among processes that
were previously isolated from each other in time or space. Information can be understood as a mediator of interaction. Decreasing the
costs of its propagation and storage inherently increases possibilities
for interaction effects. An Information Revolution is therefore likely
to beget a complexity revolution.

Adaptation and Information

For many of the conditioning factors mentioned above, there is often an adaptive mechanism buried in their inner workings. Some adaptive mechanisms are simple and work without agents' being aware of consequences for others. An example is the network externalities of fax machines, where each new machine makes all machines more valuable by increasing the pool of others to which they can connect. Once enough users have purchased fax machines, the spread of fax becomes self-reinforcing. Other mechanisms are elegant accomplishments of human intellect, such as the world propagation of easy computer cryptography systems by members of subcultures intent on fostering individual liberty at the expense of government potency.

Adaptive interactions are, in fact, a major raison d'être of the Information Revolution. Improvements in processing, storage, transmission, and sensing make it possible for us to know the state of a system with far greater speed and precision. We want this knowledge because it allows us to be more adaptive, and that in turn can vastly increase performance. Antilock brakes allow adaptation to road conditions at a time scale faster than native human capabilities permit. Financial networks allow buying and selling based on global knowledge of price movements that could not earlier be assembled. Effects of military attacks can be known from sensors and satellites, allowing adjustments in later attacks. Effects of policies in business and government can be assessed much more accurately and quickly, allowing for adjustments to policies (such as monetary rates, inventory acquisitions, or licenses of new pharmaceuticals) that were unthinkable in previous generations.

Much of the promise of the Information Revolution rests on the possibility of increasing the pace of adaptation in our (often complex) social and technical systems. It is ironic that exploiting the promise of short-run possibilities for better prediction and control (such as linking financial markets) can create longer-run difficulties of prediction and control (such as global propagation of financial crises). But the cumulative effects are clear. The exploitation of new information technology to create desirable adaptation increases the linkages that foster systemic complexity. Some variety is lost in the standardization of protocols that is needed for effective communication. The gain in the breadth and depth of interaction that results allows a large diversity of actors to be part of the same Complex Adaptive System, thereby increasing the opportunities for adaptation and the level of interdependence.

The Information Revolution engenders Complex Adaptive Systems for reasons that we can now see are intrinsic. To secure the benefits (and avoid the pitfalls) of this enormous change, designers of every kind of enterprise, public or private, need a framework that captures the fundamental relationships of information to complexity and adaptation.

Complexity as a Way of Thinking

The Information Revolution not only promotes faster and wider adaptation; it can also promote a new mode of thinking about social systems. The Industrial Revolution made metaphors of machines and factory production widely available. These mechanical conceptions had a profound effect on approaches to organizational design. In business, they led to an emphasis on predictability and control. In public affairs, they led to an emphasis on rules to be executed by hierarchies of relatively expert and impartial public officials. In both settings, efficiency and consistency became preeminent goals. Of course, real processes often did not exhibit unblemished efficiency and impartiality, but these were the ideals toward which organizational activities were oriented.

Recently, there has been increasing dissatisfaction with the costs of the industrial mode of thinking and action. Its impersonality and

appreciated the need for government to protect the market from force and fraud. People now recognize a range of additional problems, including the tendency of some markets to self-organize into oligopolies or even monopolies. Moreover, information systems are subject to several modes of failure, which we discuss in the chapter on interaction. Rather than undermining the value of complexity as a way of thinking about social systems, an appreciation of how Complex Adaptive Systems can fail provides valuable guidance for the design and management of complex systems, including human organizations as well as technical systems. Designing new strategies and organizations will frequently imply altering—or even creating—the variation, interaction, and selection that are hallmarks of a Complex Adaptive System.

II

Variation

The Role of Variation

Variation provides the raw material for adaptation. But for an agent or population to take advantage of what has already been learned, some limits have to be placed on the amount of variety in the system. The key question in this chapter will be the right balance between variety and uniformity. We will examine the workings of major mechanisms that affect that balance.

When we treat a Complex Adaptive System as a population of agents, we begin by assuming that the agents are not all the same. Indeed, the *variety* within a population is a central requirement for adaptation. The surprising dynamics that occur in complex systems are often consequences of such variety, as when a long-reigning political coalition collapses with the arrival of what seemed to be a minor new participant. And the novelty or innovation that we may want to encourage will often stem from such variety, as when ideas about unmet customer needs and ideas about new technical possibilities come together in the conception of a new product.

It is often tempting to assume that the agents of a system are basically all the same—all the birds in a flock, all the employees of a particular company, all the citizens of a foreign town. Such assumptions simplify subsequent analysis. So we mass-produce kitchen tables on the implicit assumption that the population of buyers are all about the same height. Manufacturers know this makes some problems for people who actually are very tall or very short, but ignoring those issues allows the manufacturers to attain production efficiencies without fears of incompatibility with standard chairs. If there are some resulting difficulties for end users, the seriously affected individuals can respond to them with custom modifications.

Similarly, if an average family in a community has two children, it can be convenient for some analyses—say, of projected demand for schooling—to make forecasts by assuming that all families are the same. For many purposes, however, such an assumption would be a mistake. What if families of different ethnicity had different sizes? Then a change in neighborhood ethnicity could change school demand even though the number of households remained constant.

The designer of a new product, such as a video recorder or a bicycle, seeks a design that most customers will use in conventional ways that will engender no unusual problems. But to design the product, its packaging, and its instructions as though all its buyers will use it identically is generally not the best strategy. Designers need to anticipate multiple categories of use and then either target the new product to a single category or design it to meet multiple requirements.

The Complex Adaptive Systems approach, with its premise that agents are diverse, is well suited to design projects such as the video recorder and the bicycle. It builds in the default assumption that there is variety within a population that could matter. Simplifications still can be made, of course. But the issue of variation is at the forefront of the analysis rather than in the background.

The actions available to policy makers and designers who want to shape the behavior of a Complex Adaptive System often work not just by accommodating variety; they can also work by actually increasing (or decreasing) the variety of agents in the population, or the variety of product design ideas under discussion, or the ethnic variety of housing purchasers. Variety turns up repeatedly in complex systems as a crucial factor in their development. But the situation is not always so simple as saying that homogeneity is bad and variety is good.

In an ever-changing world, agents that are not currently best may be a resource for the future. Parts of them may be crucial at a later time. For example, monoculture takes a great risk by eliminating the genetic variety in a crop. Without genetic variety, the introduction of a new parasite can wreak havoc. Variation in agents may be valuable even if the environment is unchanging, if the best agents in the population up to this moment are far from the best possible. In both these cases, when the world is changing or the current agents are far from the best possible, variety can have value, and homogeneity may be a hindrance. A designer or policy maker confronting a Complex Adaptive System should therefore ask a central question: What is the right balance of variety and uniformity?

Focusing on variety in this way requires additional aids to clear thought. Real families in a community or real product users can present wildly varying blends of characteristics. Some of these matter and some do not. Which characteristics matter is partly a function of what goals are being pursued. The variable heights of video recorder buyers are presumably not consequential. But height could matter for a bicycle. Variation in ethnicity of buyers may not matter much for the design of a bicycle, but it may matter for the written instructions on how to program a new video recorder if variation in ethnicity entails differences in language.

When variety is significant, we need to be able to talk about subpopulations. We need to analyze their differences without losing track of the possibility that there are many other differences we are temporarily ignoring. Some of the potential buyers of our new video recorder may only want to play movies with it, while others might only want to record daytime shows for evening viewing. Some might be native speakers of Japanese. Some might know English as a second language. A **type** is a category of agents within the larger population who share some detectable combination of features. The notion of type facilitates the analysis of variety that our framework so often requires. Commonly, we distinguish types by some aspects of the agents' properties or behaviors that are observable, either by other agents in the population or by outside analysts. Examples of types are:

- tall, average, or short customers for a consumer product;
- individuals who have not been infected by a particular virus, who have the symptoms of the disease, or who have recovered and are immune;
- molecules with a shape that matches a receptor on the surface of a group of biological cells, and molecules lacking that shape;
- viewers of a cable weather channel who are "forecast drop-ins," "general-weather-interested," or "storm obsessed";
- invasive computer programs that are classified as "viruses," "Trojans," "droppers," or "worms."

Many type distinctions are endogenous—actively developed and used within the population by member agents themselves. In our disease example above, the general public may detect two types, the symptomatic and asymptomatic. Using these types, an individual agent may gain tremendously because its action can be conditional. Contacts with those who are obviously symptomatic can be avoided, significantly reducing the chances of acquiring the infection.

But notice that this works only if the individual can use symptoms to pick out those to be avoided while interacting with the rest. Rarely can a person make such discriminations perfectly. The symptoms a person can detect are, after all, only approximate indicators for the actual condition of other individuals. (Some with the flu may be trying valiantly to get through the day without appearing sick.) So the concept of detectably symptomatic individuals induces types in the population that are correlated with benefit to the individual, but not perfectly correlated.

The spread of the infectious microbe induces an observer of the system, such as a public health official, to subdivide the population into three types: those who have not been infected but can be, those infected, and those who have recovered and have immunity. Types are not given from on high but are defined by actors within the system.

To take another example, police would like to pick out drivers whose level of intoxication makes them dangerous on the road. But they cannot be sure of that intoxication before an accident occurs. Blood levels of alcohol cannot be observed as traffic passes by. Erratic driving can be observed, so those cars can be stopped. Some stopped cars will contain a driver who will fail a blood analysis. Some cars with a high-blood-alcohol driver will pass by. The police officer's test, erratic driving, divides the population into types, those who can be stopped and those who cannot. It predicts blood alcohol levels only imperfectly. In turn, blood levels predict dangerous driving imperfectly. For example, some drivers may perform unsafely with blood alcohol concentrations below the legal limit. Here we have a tangled set of categories used by the actors within the system. Each of the working categories (or type definitions) only approximates the issues of actual concern.

Many examples of complex systems have the property that the population contains at least some agents, such as the disease-avoiders or police officers, whose actions are conditional on aspects of the other agents that they can detect. At a border crossing the detection of a fraudulent passport can spell the difference between freedom and prison. This is an example of how a small difference related to a condition can trigger a large difference in subsequent actions. For this reason, conditional action can result in consequences that are not smoothly proportional to causes, so-called nonlinear dynamics.

So far we have used examples in which agents in the population or outside analysts employ similar distinctions to divide populations into types. Distinctions can also be made that are completely external to the population. They may be suggested because they correspond usefully to some differences in the population that matter, even though the actions of agents in the population may not necessarily be conditional on the differences. So, for example, a product designer might want to suggest that there could be two types of buyers for the proposed video recorder: perhaps "movie renters" and "time shifters." The designer need not contend that real consumers classify themselves this way. The distinction may be offered to highlight that these different kinds of users have different needs for the controls on the device. A pure movie watcher might have no need for the elaborate systems to program recording that begins and ends automatically, nor even for a clock. A pure time shifter might need those capabilities and many more. Designers might debate if these are useful types to distinguish, whether or not consumers make the proposed distinction themselves. And though real consumers might rarely fall into one of the pure categories, the distinction may help designers think about the potential market. Perhaps they will want to consider introducing a cheaper, simplified machine that only plays prerecorded films.

In the surprising world of contemporary consumer electronics, the low-end version might just be the full-featured machine with some features disabled. This can be a way of charging different types of customers different prices. Such price discrimination can raise profits even when the cheaper machine costs somewhat *more* to manufacture with turned-off features (Shapiro and Varian, 1998). Whether this will work requires designers and marketers to analyze the detailed pattern of consumer types.

Here are the five important aspects to the notion of types:

1. Types are generally defined by some detectable features of the agents in the population;
2. many other dimensions of variety in the population may persist

in the population without being recognized as types by the
agents themselves;

3. the features that distinguish types usually provide only an imperfect indicator for the actual differences in action among the
agents in the population;

4. types are often endogenous in complex systems—agents
within the population may detect types and act conditionally
(and even change type definitions if the system is adaptive);
and

5. types can be exogenous as well—defined only in the minds of
those analyzing a Complex Adaptive System from the outside.

The notion of type will help us to analyze the sources and contributions of variety by considering how systems create, destroy, and
modify types. In later chapters we will examine the way structural
factors affect the rates of interaction among types and how credit is
attributed to types as a result of good or bad performance.

Altering the Frequencies of Types

Copying With Error

We describe a complex system, whether adaptive or not, as a population of agents. The agents are instances of various possible types.
And the population has mechanisms that create, destroy, and transform the agents. Death is the most obvious transforming mechanism, destroying agents and possibly destroying a type if all its
instances die, as happened to the dinosaurs. Birth creates new
agents. Death and birth processes apply not only to biological entities but also to organizational entities such as companies and political units.

In the simplest case, copying can be understood as the most

primitive birth process. When it functions without error, the result is an increase in the frequency of one of the population types, whether the population is virus particles or documents. Copying is seldom perfect, however, although in the realm of digital technology it can now come very close.

In the genetic case, **mutation** is a copying error that serves as an important source of variety. It can function to create new types, as well as to alter the relative frequencies of existing types. It is striking that many kinds of Complex Adaptive Systems have mechanisms that function similarly to genetic mutation. For example, *temperature* in systems in which the elements have energy levels, such as the annealing of metals, also functions to "mutate" arrangements of atoms into new configurations. *Process errors* in factories and laboratories can have this same impact of creating new types. The ink-jet principle was accidentally discovered when a research laboratory syringe malfunctioned. There are many other processes that introduce "noise" into operations of copying or re-creation, thereby producing variants that are sometimes highly novel.

These mechanisms tend to have certain properties in common. They introduce variation into a system from uncontrolled forces external to the system, such as radiation, external heat, or disruptions of quality control. As a result of the uncorrelated, exogenous source of variation in types, most of the variants introduced into orderly systems by such processes are deleterious—with occasional small improvements and a sprinkling of very rare spectacular advances. Exploring for new possibilities by nearly random variation can therefore be expensive. In fact, random variation is even slower than enumerating all the possibilities, since random generation will add duplication. With random variation, you examine each piece in the haystack and put it back if it is not the needle, possibly to draw it again later.

Endogenous Copying Mechanisms

By contrast, there are a number of other mechanisms that produce new types or change in type frequencies in a more targeted, less random, fashion. They tend to be endogenous, triggered by events internal to the system in which they operate. In particular, selection creates copies of some agents or strategies from a population and eliminates copies of others.

Simple selection has an important effect. Over time, it reduces the variety of types in a finite system, although in the beginning it may increase the relative frequency of some rare types. Neither copying nor deletion generates novel types (except through errors in copying, as discussed above).

So when a personal computer manufacturer offers two models of its product, and consumers buy one enthusiastically, many new copies of the preferred design will be made. This is a kind of selection process. It will gradually result in the copies of the other design being a rare type in the product population, even though most of them may continue to function. If consumers abandon the machines of the less preferred design, their actions function like death in biological populations, reducing even further the relative frequency of that type.

When an athlete decides to greet fellow players with a "high five" instead of a traditional handshake, the choice of behavior is a form of selection, in this case selection by imitation. The agent replaces a current strategy with a new one copied from the actions of another agent—perhaps an athlete who is highly admired. No new agents are created, but one changes type. A succession of similar decisions will transform the culture of greeting in the athlete population.

In selection mechanisms there are criteria at work in determining what types are copied and what types are replaced. Consumers evaluate one version of a product as being more desirable than another version. Athletes do the same for greeting gestures. Such evaluations require some kind of attribution of credit, either explicit ("the product is highly rated by a consumer magazine") or implicit (popular athletes use the new greeting).

We will devote more attention to the attribution of credit in the chapter on selection, but now is a good time to point out the relationship between the attribution of credit and the level of variety. If the selection among types favors more common types, then a type with a slight frequency edge can grow quickly to become predominant in the population.

As we mentioned in the previous chapter, the conventional examples of this dynamic are the competition between the VHS and BetaMax systems of video recording, and the competition between the QWERTY and Dvorak keyboard arrangements (David, 1985).

Economists often refer to the benefits that accrue to an individual user from the sheer numbers of other users as **network externalities.** These examples illustrate their force in both accelerating convergence and reducing variety. When the type that is "in the lead" is the best, and conditions are not changing, rapid convergence on a standard is desirable. The convergence on a single type of recorder or keyboard can unleash considerable benefits, especially where there are strong economies of scale. There can be more movies on video and more new computers introduced as a result. But, as we pointed out earlier, in changing conditions, or when types so far available are not the best possible, the loss of variety can become a serious problem. Intensive development of the possibilities inherent in BetaMax recording or Dvorak typing does not occur. New recording formats or keyboard layouts do not survive in the marketplace, even if they might be superior alternatives.

Recombining Mechanisms

Unless copying errors are also occurring, selection only alters the relative frequencies of existing types. However, there are endogenous mechanisms that do create new types. In biology, one of the most important is **crossover,** a process of recombining genetic contributions from each of two parents. This mechanism creates novel types, but with a method vastly different from mutation. Crossover works by splicing together pieces of already viable genetic material instead of making changes at random and so is far more likely to yield an improvement than is a mutation.

It has long been noted that self-conscious activities of deliberate invention have similar properties. This can be seen in early forms of inventions, such as the motor and wagon combinations of the first "horseless carriages." In designing their aircraft, the Wright brothers defined subproblems that could be independently attacked. Solutions to the power source problem and alternative wing designs could then be recombined in various ways (Cohen, 1982). As in the biological analog, these **conceptual recombinations** have much higher chances of being valuable new types than would random changes of designs.

Constraint relaxation is another such mechanism, frequently practiced in human problem solving. It seeks solutions to a hard problem by generating variants that violate some one of the situation's constraints (Kauffman, 1995, pp. 245–71). It introduces new variants by starting with materials of established feasibility and modifying them. A nice example is the one-opening kettle, which was achieved by relaxing the constraint that kettles needed a wide hole for filling and a narrow one for pouring. Presumably, many kettle designers thought: "What if we only had one opening?" Once faucets became common for filling, instead of ladles, the two-hole constraint could be relaxed, achieving a good design that is less expensive to manufacture.

In crossover, conceptual recombination, and constraint relaxation, we have examples of mechanisms that can both create new types and change relative frequencies. We will use the general label **recombining mechanisms** for them. Because they work with portions of strategies or agents already in use, they introduce new types by an endogenous process that has some degree of correlation with the system's other conditions. This contrasts with the exogenous sources of new types that arise from errors and random disruptions. Recombining mechanisms implicitly leverage performance criteria in their creation of new agents or strategies because they draw parts of the new agent or strategy from those that are already succeeding. This advantage over random variation explains why they are so commonly found in Complex Adaptive Systems.

Exploration Versus Exploitation

Over the years, research on these various mechanisms for creating, transforming, and destroying agents or strategies (and therefore types) has led to the establishment of an important trade-off principle, usually referred to as **exploration versus exploitation.** This principle captures the tension in Complex Adaptive Systems between creation of untested types that may be superior to what currently exists versus the copying of tested types that have so far proven best. This trade-off characterization has turned out to be illuminating across a wide range of settings from simple genetics to organizational resource allocation, wherever the testing of new types comes at some expense to realizing benefits of those already available (Holland, 1975; March, 1991). Two extremes illustrate the trade-off.

Eternal boiling occurs when the level of mutation, temperature, or noise is so high that the system remains permanently disorderly. In such a state, any potentially valuable structures are broken apart before they can be effectively put to use. Organizations and polities sometimes find themselves in periods of continuous upheaval that produce this effect. There may be striking new ideas, but before it is known whether they will actually work, their underpinnings are swept away in subsequent waves of change. If mutation rates are too high in a biological population, there will be many variant organisms that cannot survive, along with a few that are important improvements over earlier types. But a surprising new discovery will itself be subject to high rates of (usually lethal) mutation, and so its superiority will not have enough time to establish itself in the population before it is disrupted by deleterious random change. In these situations, exploration completely swamps exploitation.

Premature convergence is the opposite phenomenon. Premature convergence occurs when needed variability is lost too quickly. This can happen when very speedy imitation of an initial success cuts off future system improvements. A fashion in health care can sweep through a community and wipe out alternatives that might have proven superior with some additional testing and refinement. A new product can become a standard before sufficient trials of alternatives have occurred. A new variant that might actually have the potential to be better may lose out because it cannot overcome the advantages accruing to the established standard. In these situations, exploitation quickly swamps exploration.

These are not simply two ways that something good can go wrong. In general, investments in options and possibilities associated with "exploration" frequently come at the expense of obtaining returns on what has already been learned, "exploitation." The two possibilities form a fundamental trade-off. An early and striking exposition of the trade-off occurred in the context of the "two-armed bandit problem," in which a player with a fixed supply of coins plays two slot machines that have unknown and potentially different rates of payoff. To decrease sampling error in estimates of which machine pays more, and thereby increase longer-run expected gains, coins should be played on both machines. But to maximize gain in the short run, coins should be played on the machine that is currently estimated as best paying (Thompson, 1933; Holland, 1992).

The trade-off can be seen in many different practical situations. For example, companies must decide whether to invest resources, such as capital and management attention, in developing ideas for wholly new products or in marketing, refining—or reducing costs of—existing products. Students must decide between taking more courses in a subject in which they have done well or trying out new fields.

Though there is not a single decision maker, the gene pool of a biological species also confronts the problem. Higher mutation rates will produce more variations on currently fit animals, with attendant chances of discovering improvements. But they also will yield higher levels of infant mortality due to lethal random changes in organisms' genetic codes. The exploration of making new animals unlike their parents comes at the cost of forgoing the fitness already embodied in the parents. Lowered mutation would make offspring more similar to the parents, exploiting what the parents have proven to be valuable in reaching reproductive readiness—but at the cost of exploring less.

Example: Military Personnel Systems

The themes of exploration versus exploitation can be clearly seen in personnel policy. In the short run, it pays to promote the person who best fits the current vacancy. In the longer run, however, it may pay for an organization to sacrifice some of these short-run gains to develop a set of people who will provide a better set of options in the future. Harnessing complexity involves acting sensibly without fully understanding how the world works.

One particularly interesting example is military personnel management for the upcoming decade and more. This example shows how many of the questions we are raising come together in an actual situation where policy decisions may have vast consequences.

The period ahead has been characterized by a number of experts as likely to produce what has been called a Revolution in Military Affairs. They compare it to other periods in history when dramatic changes occurred in military capability and doctrine. It is evident to military planners that future combat could be very different as a result of advances not only in information technology but also in nano- and bio- technologies.

Of course, it is not possible to say now just what those differences will be. While technical changes of the current magnitude have historically had major implications, the path from technical change to effective military operational capabilities is often convoluted, and the actual implications are frequently not what was anticipated. For example, it took years to discover the best use of the advantages conferred by night-vision equipment and stabilized naval gunnery platforms. The early prospect for aircraft seemed to be reconnaissance (Burton, 1997). The arquebus, an early gun, is a typical case. It took nearly fifty years before it provided an important military advantage. Though the gun was powerful against archers, it was slow to reload. It could only be effective after a drill called countermarch was invented and taught to soldiers. Then a rank could advance and fire while other ranks fell back to reload (McNeill, 1982). The technology alone was not effective. Routines had to be found, refined, and disseminated to make it effective, and this required decades to accomplish.

A key observation is that the transformation of technical possibilities into meaningful capabilities is frequently accomplished concurrently with the career developments of personnel. This is true for all kinds of organizations, not just armies. It suggests that employees of organizations or officers in military services might be usefully analyzed as populations of agents that form a Complex Adaptive System.

In the military case, younger officers become interested in new possibilities. They find opportunities to experiment. Some of them are supported and protected by more senior officers who are sympathetic. As their careers advance, they get continuing opportunities to refine their ideas on things like naval aviation, tank tactics, or airborne logistics. And their careers give them access to increasing resources for testing their ideas in field exercises and war games. At some point combat situations arise in which some of these ideas are put to the most stringent tests. Eventually, some of these idea-bearing officers may be able to forge an entire career system for other officers who want to pursue some new mode of combat (Rosen, 1991). This is what happened to Admiral William Moffett, whose career became synonymous with the rise of naval aviation. It is also the story of Admiral Hyman Rickover and the development of nuclear submarine forces, although both officers entered into the new area midway through their careers.

The problem facing today's makers of military personnel policy is that no one can know which young officers have the key ideas for the surprising new operational concepts of the next decade. By definition, the big surprises are those deviating enough from incremental change that they cannot be confidently foreseen. As a result, the specialists within the military services who make personnel decisions face precisely the kind of hard-to-predict situation that is common in Complex Adaptive Systems. The British, for example, could not foresee that merging their early naval flyers into the newly formed Royal Air Force, which was dominated by long-range bomber pilots, would seriously interfere with efforts to implement technical innovations that could apply to aircraft carriers (Rosen, 1991). While most of the technical advances required for carrier combat were made in British experiments, those new technologies and methods ended up being fully exploited by American naval flyers and commanders. The loss of the naval career path for experienced British naval pilots is a major part of the story of why the British could not implement their own innovations.

Several of the principles we have come to in our review of complex systems research are relevant here. Two considerations suggest that now is a good time to weight exploration heavily relative to exploitation for military personnel systems. First, many military experts believe that the current profusion of technical possibilities, including the Information Revolution, makes radical changes in combat over the next two decades seem extremely likely. Second, for the United States in the immediate future, there is no military enemy of comparable strength.

If we accept the premise that exploration is to be strongly emphasized, the Complex Adaptive System perspective suggests several questions to ask about how to pursue that goal. First, what are the sources of "new types" of strategy-bearing officers and changes in frequencies of types? It seems wise to expose junior officers to new technology and its possibilities at the earliest possible points in their training. It may also make sense to include the biographies of previously innovative officers in their training, to help them see that their own vision of future possibilities can be linked to their career prospects. The concepts that will be vital in the future involve combining knowledge of military requirements with knowledge of new technical possibilities. Officers will be valuable who have mixed exposures to the two knowledge bases, and who appreciate how previous innovators managed their careers.

The personnel example lets us foreshadow some of the issues that will be discussed in the following chapter on interaction patterns. If the aim is to produce officers with unusual and useful combinations of knowledge, what controls the interactions among types? The system that assigns new posts to officers takes on central import. To the extent possible, the assignment system should mix experiences with the deliberate intention of exploring new technical possibilities and operational concepts. It should accept more "risky proposals," whose payoff may not be obvious. It should also create crosscutting contact networks for future use in allowing easy recombination.

This means that the apparently boring personnel activity of rotating officers through posts needs to be deeply informed by the best recent theories of new technology and related combat possibilities. Of course, realism dictates that many other factors must play a role in such placements. But considerable benefit can be gained from every step that can be taken to increase the career controllers' vision of the "space of possibilities" being explored by the individuals in the organization as they develop their careers.

What other barriers should be erected or removed in order to change patterns of interaction among types? For example, policies that keep assignments within interest areas will affect who interacts with whom. In our military example, if an officer develops an interest in logistics, is it better for the next placement to be in logistics? Is it the same for all types? Should the rates of rotation be similar for most officers, or should they be a function of the outcomes of previous rotations, with some officers therefore having many shorter postings?

Finally, we can observe that the officers will be evaluated as their careers progress. At each stage of their careers, the officers, and therefore the new ideas they embody, need to be assessed, and the character of this assessment will determine the speed and direction of the system's adaptation. This leads to many questions about the attribution of credit, which will be discussed in Chapter IV, on selection. For example, the officers can be expected to adjust to the rules established by personnel policy makers. Some of those adaptations may be undesirable and cause the policy makers to make further changes. The system may do better overall if the policy makers, who must avoid making changes too rapidly or slowly, can identify the coevolutionary character of this dynamic.

Thus a Complex Adaptive Systems perspective generates a systematically interrelated set of questions about the personnel management area. These are questions that arise, as we noted at the outset, because Complex Adaptive System theory suggests general analogs for the principles of artificial selection that once fascinated Darwin.

Whether to Encourage Variety

The preceding example began with the premise that encouraging variety might be the right thing to do. But often the problem facing designers or policy makers is not *how* to foster variety, but *whether* to do so.

As with nearly all aspects of Complex Adaptive Systems, we know of no general rules for when one would reliably do better with more variety. You have to analyze actual situations, and then you have to place your bets. But in the course of studying many Complex Adaptive Systems, we have observed some broad factors that seem to reduce the costs or increase the benefits of exploring relative to exploiting. If you are in an ideal situation where you are sure your current approach to a problem is the best that is possible, and you do not think the problem is going to change, then any exploratory deviation from it should be avoided. But for more realistic situations, we can identify some conditions in which exploration is especially valuable.

1. *Problems that are long-term or widespread.* The more use you can make of an improvement, the more it pays to bear the costs of searching for one. For example, if someone buys a house in your neighborhood, it may well be a good investment to take some exploratory actions that risk not being reciprocated but could establish cooperation that would be valuable in the long run. Likewise, a problem that occurs in many branch offices of a company may provide a good opportunity to try local experiments if progress in one branch could be made widely avail-

able for adoption at the other branches. When problems are long-term or widespread, there may well be good bang for the exploratory buck.

2. *Problems that provide fast, reliable feedback.* This is a closely related condition. If you can learn quickly whether an alternative solution might be better, and if there is not a big chance of being misled, then you have more chances to find an improvement. Moreover, you have longer to gain from what you might discover—and you are more sure that what seemed to be an improvement actually is one. Where such fast and accurate feedback channels don't exist, it is often worth trying to create them so that the benefits of exploration can be gained. In the sciences, which are heavily oriented toward exploratory activities, the areas that attract the most effort tend to be the ones in which new experiments or observations can be done rapidly and in which repetitions of prior studies give similar results. Experimental psychology grows faster than anthropology. Similarly, companies that can learn quickly about consumer reactions can afford to explore more of the space of possible products.

3. *Problems with low risk of catastrophe from exploration.* In some situations, you may be able to judge that the risk of an extremely bad result from exploration is low. That should increase the amount you are willing to do. In training sessions, gymnastic competitors can exploit current capabilities by practicing elements they already know. These can always be done a little better, and repetition is needed to avoid mistakes during competition. But gymnasts also need to attempt new, and more difficult, moves—ones they have never done before, occasionally even ones that no one has done before. This is exploratory activity. The new elements may never be mastered, and in the meantime old ones have not been practiced. There is also increased risk of injuries that could end a season or a career. Practice spaces are equipped with facilities to reduce injuries,

such as extra padding for apparatus and ceiling-suspended safety harnesses. These limit the risks of serious injury, though they do not eliminate them. The practice safety devices are installed exactly so that the costs of exploration will be lower. Inventing such devices greatly accelerates the evolution of otherwise dangerous activities, be they sports or flight maneuvers, by transforming the situation into one with lower risks of catastrophe. Another class of devices for lowering the risk of catastrophe turns on having multiple replications, with the fate of the whole system not resting on any one. Thus a federal system can prosper when many states act as partially independent "policy laboratories." A drug company can use cell cultures to test a vast array of chemicals, most of which will be ineffective, in searching for a few that might be useful.

4. *Problems that have looming disasters.* If continuing to exploit the best solution found to date is apparently going to lead to disaster, then of course one is wise to explore. Everyone is familiar with the notion of a desperate gamble on a novel approach, such as a daring attempt to escape from death row. Our argument for exploring when catastrophe seems imminent may seem almost a contradiction of our previous arguments for exploring when there are long time horizons or low risks. But in this situation the relative attractiveness of exploring comes from the negative yield of exploiting. The results of exploring and exploiting are measured on scales that have real zero points. As we will see in discussing extinction: not only can you do worse, you can go completely out of business.

Example: Linux Software Development

We can illustrate the conditions that favor encouraging variety by considering the striking case of the Linux computer operating system and the method used to organize the work of its developers. The method is known as open source software development. This form of software development has been thrown into the limelight by the spectacular growth of Linux, which has become, in certain key areas of application, a serious competitor to operating systems developed and sold by major corporations such as Microsoft, Sun, and IBM (Moody, 1998).

This is a very surprising turn of events since Linux is given away free by its developers. There is a natural presumption that free software cannot be as reliable as for-profit software. Yet it is precisely for situations demanding high reliability that Linux has found its strongest support.

The surprise deepens with the observation that Linux is not only free but also the handiwork of an enormous, worldwide cadre of unpaid volunteers. By some estimates, Linux is the result of contributions from many thousands of programmers. A computer operating system is one of the most intricate of human creations. This number of cooks would seem more than sufficient to spoil the soup. How could thousands of scattered volunteers make an operating system that is more reliable (and faster running) than those created by dozens, or hundreds, of highly talented programmers working full-time for renowned corporations?

Considering the development of Linux as a Complex Adaptive System casts light on some important components of the explanation. We can begin by pointing out that Linux is not the only example of the open source approach to software development. There are many earlier examples, such as the scripting language Perl and the e-mail server sendmail. The most widely deployed software for serving up requested World Wide Web pages, a system known as Apache, is also the product of volunteers working together in an open source framework. What all the examples have in common is the free availability of the source code, the human readable computer instructions that specify the program. That arrangement provides the generic label for this approach to team software creation: open source software development.

The free access to the source code of Linux means that any programmer with sufficient motivation can craft changes to the code, creating a new version of the program. This is not possible in traditional development with proprietary code. From a Complex Adaptive Systems point of view, the possibility for volunteers to create working variants increases massively the variety of the population of operating systems. In successful open source cases such as Linux, that variety has been harnessed to yield a very effective result, although many observers expected chaos to result from the rapid injection of many potentially incompatible variants.

Our framework points to several structural arrangements that work to make the added exploration beneficial, averting the prospect of death by eternal boiling. In our terms, when a programmer modifies the source code of Linux, this activity is an endogenously triggered recombination. The trigger is usually an observation of some particular kind of poor performance by the existing standard version of the operating system. The affected user may make an electronic request for help from the large Linux community. Interested individuals respond by suggesting fixes. These small pieces of new code are recombined with the rest of the standard version to produce new variant versions. A period of testing and discussion of the performance of the variants follows. Eventually the best-performing variant is accepted by the small team of key Linux developers, who incorporate the new code into a subsequent standard version of Linux.

When open source development prospers, a central reason seems to be that "given enough eyeballs, all bugs are shallow" (Raymond, 1998). For Linux, there are certainly enough volunteers. Equally important, the communication among volunteers about triggering problems and proposed alternatives is precise enough that multiple plausible variants are routinely generated as possible solutions to most problems that bother users. In addition, testing of alternatives is reliable enough that the code that wins out is generally very good code, with unwanted side effects being rare. Thus the variety made possible by the free availability is marshaled to produce a rapid rate of improvement in overall quality. By inquiring a bit further into how this is accomplished, we can uncover some clues about when an open source approach is likely to work well—and when it is not.

A crucial fact is that there are two types of Linux versions: standard and variant. The few central managers of the Linux community, led by the originator of the operating system, Linus Torvalds, retain the right to label versions of the system as official releases. Each new official release creates another "standard Linux," and millions of digitally perfect copies are made of it.

This control over the definition of the next generation of the operating system is strikingly analogous to a biological mechanism seen in the emergence of multicellular organisms: sequestration of the germ line. This is a restriction of reproductive activity to a few specialized cells, while the vast majority of cells in the organism no longer participate in creating the next generation (Buss, 1987; Maynard Smith and Szathmary, 1999). In both cases, limiting "reproduction" to a tiny fraction of all the agents reduces the chaotic inconsistency that would follow if all variants had equal opportunity to shape the future.

In the Linux case, the centralized control of changes in the standard code makes higher levels of variety in the proposed changes sustainable, so that the "law of sufficient eyeballs" can come advantageously into play. In the biological case, the restriction functions to alter the evolutionary "incentives" of cells making up the organism. Over succeeding generations their strategies will be far more likely to be those that let them prosper as a "team" rather than those that benefit individual cells at the expense of the others (Dawkins, 1989). Analogous incentives are created for the programmers in the Linux case.

Numerous experiments are being undertaken in an effort to imitate the striking success that the open source approach achieved in the Linux case. As usual, we do not claim to be able to predict the success and failure of particular efforts. But Complex Adaptive Systems principles do suggest a number of key questions to ask when contemplating an open source software project. Several of these come from the preceding section on whether to encourage variety. As we have seen, variety is the engine of rapid quality improvement in an open source initiative.

We can see that Linux has at least three, and perhaps all four, of the conditions favoring exploration that we outlined earlier.

1. *Problems that are long-term or widespread.* In contrast to computer hardware and applications programs (such as Web browsers), operating systems are among the longest living elements of the computational world. Unix—of which Linux is a free version—dates back to 1969. It runs on mainframe and minicomputer architectures that long predate the microcomputers that now cover the earth. Thus, an improvement to an operating system is likely to bear fruit over a very long period (as time is measured in the strange universe of computing, with its Moore's law of doubled computer power every eighteen months). Another example of open source development, the Apache Web server, also seems to occupy a functional niche where improvements can be expected to have long service. In addition, the gains from any improvement in a standard version of an operating system can benefit thousands or even millions of users, providing widespread benefits.

2. *Problems that provide fast, reliable feedback.* Linux exhibits this characteristic as well. In its typical role in server environments, its features are exercised at very high rates,

and defects become evident quickly. Moreover, open source distribution means that every contributor of a proposed variant can make a completely functional new version that can be tested locally. This further increases the rate of feedback. And finally, the quality of proposed variants can be assessed with relatively high reliability. Speed of operation and resistance to crashes are highly valued criteria across the entire community of Linux developers. Disagreements do occur over how these should be measured, and other criteria are also important. But when compared with other software areas, such as user interface design, the appropriate performance metrics are relatively clear.

3. *Problems with low risk of catastrophe from exploration.* Various parts of a software system have high levels of interdependence. When there is a premium on speed, as there is for many operating systems, there is a strong temptation to increase even further the interdependencies among modules. This can create a substantial risk of catastrophe. But Unix, from which Linux derives, has long been a partial exception to this tendency. In the Unix/Linux culture there is a well-developed philosophy of modular isolation. A key component, called the kernel, is optimized for speed. But the numerous other components are expected to honor a different set of constraints. There, interdependence among components is governed by strict principles of modularization that severely limit side effects that any activity might have on other activities. Speed is also important outside the kernel but has to be found within the architectural constraints that give primacy to crash resistance, thus lowering the chances of catastrophic consequences from exploration.

4. *Problems that have looming disasters.* This last factor favoring exploration is not a property of Linux open software development but rather of the motivation of some of the developers. Among those who have made major contributions

to Linux are many who feared the extinction of the Unix operating system family, in which they have invested their expertise. They also feared the rising hegemony of operating systems from Microsoft Corporation. For them, joining a relatively high-risk, exploration-maximizing software project may have been an attractive alternative to domination by what they often call "The Beast from Redmond."

Taken together, our four conditions show Linux to be a development project for which it is highly promising to strongly encourage variety. It does not follow that open availability of source code is a form of magic that will cause all software projects to prosper as Linux has. Indeed, our analysis suggests that in order for the decentralized generation of proposals to be effective for Linux, several other conditions were important. In particular, Linux development benefited from the ability to identify specific problems, make accurate copies of the current system with only deliberately introduced changes, evaluate the effectiveness of proposed solutions, and centrally control the choice of which proposals are implemented as changes in the standard version. It remains to be seen just how widely applicable the decentralized generation of alternatives can be. But open source software development clearly demonstrates that even very large and highly structured systems, like Linux, can benefit from the encouragement of variety.

Extinction—The Vanishing of Types

This brings us to a final set observations about the creation and destruction of variety. Typically, real populations have finite numbers of discrete agents. This means that the destruction of agents or strategies can result in the complete loss of all instances of some type.

In many conventional theories for analyzing populations this is not so. Those theories are often based on assumptions of continuous variables. Some complexity researchers have taken to calling this the nano-fox property, after conventional predator-prey models that have continuous numbers of animals in them, growing and shrinking by proportionality constants. In such conventional models a tiny fraction of each animal type is always around, so that no matter how severe the starvation, the predator population will rebound as soon as prey return. There is no complete extinction in such models. A nano-fox is always lurking in the shadows.

But in real populations the difference between having a few animals and zero animals is usually not just a little extra waiting time. Re-creating a lost type is very unlikely, and occupation of the vacant ecological niche by another species is far more to be expected. Because Complex Adaptive Systems researchers are especially interested in variety, they often use modeling tools that allow genuine extinction.

This is much more than a minor difference about the technical tools of systems modeling. The tools embody widespread habits of thinking about variation. A habit of ignoring the sharp effect of an extinction is inconsistent with many important social and policy settings. The legal system distinguishes death from the most severe and permanent incapacitation. Bankruptcy has quite different effects on a firm's history than mere extreme debt. These "zero points" in social situations correspond to sharp changes in the later dynamics. Compare what can happen before or after the last speaker of a language dies or the last copy of an ancient manuscript is lost. Our way of thinking about variety has to capture these compelling aspects of social life.

A related notion in continuous modeling traditions is that all possible types already exist in tiny quantities. This is akin to Plato's notion of discovery as a form of remembering truths already dimly known. Again, the Complex Adaptive Systems view is based on a discrete view of events and entities. It therefore suggests that a new idea may not simply be waiting in the wings for the circumstances that will bring it rapidly to prominence. It matters enormously whether the number of people who have thought of it is one or zero. We see this when we observe, once a theorem is known to be true, how readily theorists obtain the second and subsequent—shorter and more elegant—proofs. The distinction has relevance for policy strategies such as "counting on the market to find a solution," which can be expected to work far more rapidly and reliably in domains where several approaches have been partially worked out, as opposed to domains in which a feasible approach is yet to be conceived.

The underlying source of this sharp effect of zero is that copying mechanisms work quite differently from mechanisms that recombine types in context. Consider human imitation. A copying process can rapidly spread an existing type, such as double-entry bookkeeping or the Grameen banking system for microcredit. But simple imitation—even with random error—will only very rarely invent new types where they have never existed. Much more commonly, such innovations are the result of processes like crossover or constraint relaxation. A type may spread rapidly by copying once an instance of it exists, but the time required for mechanisms to create an instance of a particular type can be very long. While reproduction can be quick, creation may require a long wait. This argument about the special value of rare types has become very visible in recent years through debates about preserving plant species for future medicinal discoveries. But it has much wider application to areas as diverse as incubation of small businesses and preservation of skills that are vanishing in the Information Revolution.

• • •

Variety plays a crucial role in Complex Adaptive Systems. Our focus in this chapter has been on mechanisms that create and destroy variety in populations of agents. These are fundamental to the way populations of agents change their composition over time. They include *simple copying, copying with the introduction of errors,* and *recombining mechanisms* that create new types by systematically reusing or modifying old ones. The notion of *type* makes it easier to discuss the categories of which agents are instances and the way that the mechanisms alter a population's variety.

We have shown that there is a fundamental trade-off between *exploitation* and *exploration,* between creating agents or strategies similar to types that have already been successful and creating agents or strategies that are likely to be substantially novel. We examined four factors that may reduce the relative cost of exploration: long time horizons, fast and reliable feedback, low risk of catastrophe from exploration, and danger in the status quo.

Finally, extinction of types is especially significant in Complex Adaptive Systems. Shortly after its creation, a good new type of agent or strategy may be a rarity. Chance events when frequencies are low can have large effects, as valuable rare types (new or old) can be lost, or amplified into a more secure and effective role in the population.

We have examined the way that new types can arise among agents or strategies. We are now ready to consider patterns of agent interaction and how they influence the development of agent populations. Who interacts with whom in a complex system, and what difference does that pattern of interactions make?

III

Interaction

The Importance of Interaction

In the previous chapter, we saw how a population approach to complex systems leads naturally to thinking about the variety of agents and their strategies, and hence to the question of the right balance between variety and uniformity. This chapter is centered on the second of our three key questions: "What should interact with what, and when?"

This question also leads to ideas for specific mechanisms to harness complexity. The mechanisms dealing with interactions fit conveniently into two classes: external and internal. The external mechanisms are ways to modify the system from the outside—for example, by designing artifacts, or by policy making that changes the rules others play by. The internal mechanisms are ways to change the interaction patterns that are driven by processes within the system.

Interaction is essential to our framework because the events of interest within a system arise from the interactions of its agents with each other and with artifacts. Trades occur when buyer meets seller. Strategies of bidding and offering are enacted. Goods change hands. New animals are created when a male and female breed. Religious communities grow as adherents proselytize the uninitiated, spreading their strategy to convert other types to their own. Interaction patterns shape the events in which we are directly interested (such as trading), and they provide the opportunity for the spreading and recombining of types that are so important in creating (and destroying) variety. Interactions make a Complex Adaptive System come alive. The system becomes not a mere pile of agents of varying types but a population that gives rise to events and has an unfolding history. Those events drive processes of selection and amplification that ultimately change the frequency and variety of agent types, as we will explore in the next chapter. Interaction patterns help determine what will be successful for the agents and the system, and this in turn helps shape the dynamics of the interaction patterns themselves.

Most Complex Adaptive Systems have distinctive interaction patterns. These patterns are neither random nor completely structured. Here are two examples.

- A leader may have the opportunity to broadcast messages simultaneously to many others, who usually do not have as much capability to broadcast back. The pattern of interactions is highly asymmetric, very different from one where each agent interacts equally with all others.
- It is convenient to shop in stores near our homes. Schools and churches are often in our neighborhoods. In all these places, we meet new people. As a result, our network of acquaintances has a strong local bias. We know many people near where we live or work, and only a minuscule proportion of others in the world. Again, the pattern is far from uniform.

It is surprising in considering these everyday examples that so few tools are available to help understand the effects that flow from nonuniform patterns of interaction. A major contribution of research on Complex Adaptive Systems has been to develop ideas that help us see the sources and consequences of distinctive (nonuniform) interaction patterns.

The next section sets out three useful concepts for understanding how interaction works: proximity, activation, and space. With these we will be ready in the following sections to examine mechanisms that can harness complexity by altering patterns of interaction. To indicate the value of these concepts and mechanisms, here is an extended example of how the patterns of interaction can contribute to establishing trust and cooperation in economic and social activities.

Example: Social Capital

Robert Putnam (1993a) has found that patterns of social interaction can help explain why some communities function much more effectively than others do. Because he explains his findings so well, we closely paraphrase his summary account (Putnam, 1993b). Based on his extensive research in diverse regions of Italy, the key factor is "social capital," the features of social organization, such as networks, norms, and trust, that facilitate coordination and cooperation. Social capital enhances the benefits of investment in physical and human capital. Working together is easier in a community or organization blessed with a substantial stock of social capital.

Putnam investigated Italian regional governments to see why some are successful and some are not. He found that in successful regions citizens are engaged by public issues, not by patronage. They trust one another to act fairly and obey the law. Social and political networks are organized horizontally, not hierarchically. These "civic communities" value solidarity, civic participation, and integrity. In unsuccessful regions, the very concept of citizen is stunted. There is little engagement in social and cultural associations. From the point of view of the inhabitants, public affairs are someone else's business—the "bosses" or the "politicians." Laws, almost everyone feels, are made to be broken, but, fearing others' lawlessness, everyone demands sterner discipline. Trapped in these interlocking vicious circles, nearly everyone feels powerless and exploited.

The historical roots of these differences are astonishingly deep in Italy. Enduring traditions of civic involvement and social solidarity can be traced back nearly a thousand years, to the eleventh century, when communal republics were established in places like Florence, Bologna, and Genoa. These are precisely the same communities that today enjoy civic engagement and successful government. At the center of this civic heritage are rich networks of organized reciprocity and civic solidarity—guilds, religious fraternities, and "tower societies" for self-defense in the medieval communes. In the twentieth century, there are cooperatives, mutual aid societies, neighborhood associations, and choral societies. These horizontal civic bonds have supported levels of economic and institutional performance generally much higher than in the South, where social and political relationships have been vertically structured.

These communities did not become civic simply because they were rich. The historical record strongly suggests precisely the opposite: they have become rich because they were civic. The social capital embodied in norms and networks of civic engagement seems to be a prerequisite for economic development, as well as for effective government.

Social capital supports good government and economic progress in several ways, Putnam found. First, networks of civic engagement foster sturdy norms of generalized reciprocity: I'll do this for you now in the expectation that down the road you or someone else will return the favor. A society that relies on generalized reciprocity is more efficient than a distrustful society, just as money is more efficient than barter. Put simply, trust lubricates social life.

Second, networks of civic engagement facilitate coordination and communication and amplify information about the trustworthiness of others. When economic and political activity is embedded in dense networks of social interaction, incentives for opportunism are reduced. Dense social ties facilitate gossip and other valuable ways of cultivating reputation—an essential foundation for trust in a complex society.

Finally, networks of civic engagement embody past success at collaboration, and serve as a cultural template for future collaboration. The civic traditions of North-Central Italy provide a historical repertoire of forms of cooperation that, having proved their worth in the past, are available for dealing with new problems of collective action.

Putnam points out that unlike conventional capital, social capital is a public good, meaning that it is not the private property of those who benefit from it. Like other public goods, from clean air to safe streets, social capital tends to be under-provided by private agents. Therefore, social capital must often be a by-product of other social activities. Social capital typically consists of ties, norms, and trust that are transferable from one social setting to another. Members of Florentine choral societies participate because they like to sing, not because their participation strengthens Tuscan social fabric. But it does.

The economic implications of social capital are illustrated by the integrated industrial districts of northern Italy. There are districts for high-fashion textiles, mini-motorbikes, and ceramic tiles, among others. They are based on small-scale, technologically advanced production. Among the distinguishing features of these decentralized but integrated industrial districts is a striking combination of competition and cooperation. Firms compete over style and efficiency, while cooperating on administrative services, raw materials purchases, financing, and research.

Putnam (1993a) shows how the findings on these industrial districts, especially by Piore and Sabel (1984) and Pyke et al. (1990), can be understood in terms of social capital. These networks of small firms have low vertical integration and high horizontal integration. They do this through extensive subcontracting and "putting out" of extra business to temporarily underemployed competitors. Industrial associations provide administrative and even financial aid, while local government provides the necessary social infrastructure and services. Norms of reciprocity and networks of civic engagement are essential for the success of these industrial districts. Networks facilitate the flow of information about such vital things as technological developments, the creditworthiness of would-be entrepreneurs, and the reliability of individual workers. Innovation relies on continual informal interaction. Social norms that forestall self-interested opportunism at the expense of community obligations arise more often here than in other areas characterized by different social networks. What is crucial about these small industrial districts is mutual trust, social cooperation, and a well-developed sense of civic duty—in short, the hallmarks of social capital.

Beyond Italy, there is a great deal of evidence that horizontal networks of informal social interaction help develop social capital, and that social capital, in turn, fosters economic growth. Examples include trust in wholesale diamond markets (Coleman, 1988), rapid formation of firms in Silicon Valley (Saxenian, 1994), dense networks of clothing manufacturers in the New York garment district (Uzzi, 1997), and social cohesion among the ethnic Chinese in Thailand (Unger, 1998). Value creation within organizations also relies on social capital as the basis for the recombining of concepts that generates ideas for new products and services (Nahapet and Ghoshal, 1998). Simple procedures such as maintaining relations with former co-workers can dramatically alter the flow of valuable business information. Actively engaging subordinates can enhance the accuracy of managers' self-perceptions (Baker, 1994).

Social capital affects not just economic activity. Effective social ties reduce neighborhood crime (Sampson et al., 1997; Kennedy et al., 1998), help students achieve their potential (Stanton-Salazar, 1997), and even increase one's life expectancy (House et al., 1988; Young and Glasgow, 1998).

Social capital illustrates how the pattern of interactions has important effects for the performance of networks of agents. We now look at just how interactions operate.

How Interaction Works

Proximity and Activation

To think more clearly about patterns of interaction, it will be useful to distinguish two classes of determinants. **Proximity** factors determine how agents come to be likely to interact with each other. **Activation** factors determine the sequencing of their activity. The distinction, with good reason, roughly generalizes that between space and time.

The term "proximity" focuses attention on the many factors that make particular agents likely to interact. The most obvious of these factors is the physical space in which buyers and sellers, frogs and flies, Democrats and Republicans, friend and foe, all play out their lives. Nearby location in two-dimensional or three-dimensional physical space makes interaction events more likely for a wide range of processes, from pollination and friendship formation to predation and enemy formation.

Physical proximity is not the only kind of proximity. Normally, we pay less attention to a host of other relational networks that establish proximity, such as organizational hierarchies, old friendship ties, or community group affiliations. But these factors also determine which agents are likely to interact, and thus profoundly influence the spread of rumor and disease, the finding of jobs and marriage partners, and the occurrence of crimes and kindnesses. As the technologies of information advance, the workings of these factors become ever less tied to physical space. Friendships can be sustained by long-distance telephone calls. Communities of common interest can form through the World Wide Web.

This sampler of proximity factors has mostly been discussed as a set of static forms within which fast processes play out—hunting prey in physical space, or finding jobs in friendship networks. But Complex Adaptive Systems research often shows that on larger time scales the relationship can be reversed. In the short run, neighborhoods shape the choices of house buyers, but housing purchases ultimately shape neighborhoods. A structure that seems fixed in a short time frame may be changeable in a longer one. Here again we have coevolutionary dynamics.

Just as with movement that alters spatial distance, so most of the other proximity factors mentioned have associated change processes. Functional relationships in business are reorganized to move some groups closer together and to move others farther apart (whether or not their offices are moved). Friendship links form and dissolve. Community groups are joined and left, formed and disbanded. Barriers and boundaries are deliberately introduced into systems (physical and social) with the aim of altering the rates of interaction among types.

The term "activation" groups together many different processes that affect the timing of agent activity. Just as many different factors can be analogs of physical distance in determining interaction likelihoods, so many factors can alter the temporal structure of events. Complex Adaptive Systems research often shows that it is valuable to distinguish systems with externally "clocked" activations, such as budget cycles or seasonally triggered agricultural processes, from internally activated processes in which the results of the current event control which events may next occur. An example of an externally activated system is John Conway's famous computer simulation known as the Game of Life (Poundstone, 1985). The simulation produces its striking patterns only when all the agents (Conway's "cells") act in simultaneous lockstep (Page, 1997). Some examples of internal activation are: the movement of a sand grain in a pile that makes other grains more likely to move; the activation of a neuron that makes other neurons more likely to reach their activation thresholds; and the mobilization activities of a citizen, making those who are socially proximate more likely to become active.

The difference between external and internal activation processes can be profound. Markets where every actor can trade one unit per session will work very differently from markets where the actors with the strongest demands can trade much more frequently than others (Page, 1997). In nonmonogamous biological populations, females often follow the once-per-time-period principle (based on time required for pregnancy), while the activity of males may be limited only by mating opportunities. By virtue of this difference in their activation processes, females and males thus have quite different impacts on the composition of subsequent generations. A particularly fit male may have many matings and therefore very many copies of its genes in the next generation. A fit female in contrast will transmit its genes to a rather smaller number of additionally surviving offspring. Such biological systems are striking in the way they simultaneously make use of the intense and diffuse interaction modes we will describe below.

In designing a Complex Adaptive System, there is often some freedom to assign powers of activation internally, to individual agents acting locally, or externally, controlling activation more globally. It is the difference between "fire at will" and "ready, aim, fire." In Anglo-American intellectual traditions, decentralization is normally assumed to be an advantage. It is typical to expect the adaptive capacity of a system—especially a firm or market—to be increased when events can be activated locally and flexibly rather than globally and rigidly. But it is essential to point out that adaptive capacity is *two-edged*. As we saw in the simple case of population effects of organism death, adaptive capacity can speed extinction as well as increase viability. Allowing financial traders to respond to local conditions can let them quickly exploit short-lived arbitrage possibilities. But when globally determined prices contradict traders' assumptions, it can also let them make a rapid sequence of ever-riskier trades to cover their own losses. In our chapter on variation, we found that exploration is not always preferable. Similarly, we see here that neither greater internal control over activation nor higher activation rates are necessarily better.

Once again there is an important trade-off principle inherent in these observations about interaction patterns. It is not identical to "explore versus exploit," but it has a similar flavor. Where structural arrangements affecting proximity or activation are designed or analyzed, a major question is whether interactions will be concentrated among a few pairs of types or will be spread across a wide range of type pairings. The interactions might be accomplishing any mix of exploring and exploiting, which is why the trade-off involved is not identical. What is involved is rather the trade-off between intense versus diffuse interactions among types. Over time, are the interactions of an agent repeatedly with others from a limited number of types, or with others drawn from a wider range of types?

For example, in many countries children stay together in stable groups and keep the same teacher as they move through elementary grades. In North America, by contrast, children tend to have new teachers each year. Where schools are large, the groupings of children are also shuffled. Comparatively speaking, children in the former system have what we are calling intense interaction patterns. The other children and teachers are the same for many years. The latter system is more diffuse, with new children and teachers entering a child's life each year.

The concern that commonly arises in the schooling system with intense interaction is about insufficient exploration and loss of variety. Children and teachers may become stuck in their ways. The frequent concern about the diffuse system is that prior accomplishments and strengths may not be fully exploited in subsequent classes with new teachers and schoolmates. But there is nothing inherent about this alignment. *Diffuseness* of type interactions can also favor exploitation. This is what happens in our example of nonmonogamous males, who interact with many females, allowing the population to exploit the advantages of their genes. The point about the intense/diffuse trade-off is that it alerts us to a set of questions that need to be asked about how the channeling of proximity and activation in a Complex Adaptive System will affect the exploration-exploitation balance, along with other aspects of the system. Those questions are very important to ask, but the answers must be tailored to the specific circumstances.

These trade-offs are fundamental to "edge of chaos" arguments that have received wide attention. Their underlying claim is that evolutionary systems tend to structure diffuseness of their interaction patterns to achieve a good balance between exploration and exploitation. A typical example of such arguments is the work of Stuart Kauffman (1993), positing that evolutionary processes adjust what we are calling intensity of proximity and activation so that systems are likely to avoid both "premature convergence" and "eternal boiling." The "edge of chaos" claim has been much debated (Mitchell et al., 1993), but the debate is whether some parts of nature tend to a particular balance in the trade-offs we have described, not whether the trade-offs exist. Kauffman believes that systems tuned to a favorable balance between exploration and exploitation will tend differentially to survive. This notion of differential survival raises a set of fundamental issues we will discuss in the next chapter. Here our central question is how the channeling of interactions affects the copying and recombining of the types within a population.

Spaces: Physical and Conceptual

It is useful to examine a fundamental property of agents, the fact that they are located in space and time. When they interact, they are either co-located, or they interact via technology which is itself located. So interactions also can be said to be located. "I heard about it at the town square." "Please call me at my daytime phone number." "He bought it from a mail-order catalog house."

It follows that the movement of agents in physical space and time changes their proximity, and this in turn affects their ease of interaction. ("I'll be in my office tomorrow morning. Can you drop by then?") Moreover, actions that alter possibilities for movement in space and time will alter proximity. ("I cannot make the 10 A.M. meeting, because the airline we are required to use does not have an early flight.")

So far, we are considering how interaction patterns are affected by physical time and space: the coordinates of latitude, longitude, altitude, and Greenwich mean time that a precise global positioning device can read out. Indeed, the Information Revolution will bring many more artifacts into our futures designed to "know where they are" in space-time. Already some cars and computers—and maybe soon your portable telephone—can obtain their location by means of global positioning signals from satellites. Your strategies, and theirs, can take their locations into account, thereby changing patterns of interaction.

While our discussion began from the idea of physical space, we can use the idea as an analogy and talk about the location of agents and interactions in conceptual spaces as well. For example, an organization chart provides a map of a conceptual space. A person may be appointed director of purchasing. This is a definite location in a company's hierarchy of job responsibilities. It places the occupant of the job "near" the people who do purchasing, in the sense that these people are likely to interact with the director. Their proximity is increased. They may be nearby in organizational space even if the purchasing officers are distributed around the world and do not have offices at the headquarters where the director sits. At the same time, the organizational structure may place the director "far" from someone working in marketing, although that office is just one floor down in the headquarters building. The logic of their two roles in the business may make them less likely to interact.

The weekly senior staff meeting in this purchasing organization is thus a location in conceptual space and time. Moving up the organizational job hierarchy is a movement in conceptual space and corresponds to changes in interaction patterns. The patterns can change even when the promoted people keep their old offices (and hence their locations in physical space). Indeed, one useful way of thinking about organizations is as deliberately designed conceptual spaces that will "organize" the interactions of agents toward some ends.

The conceptual spaces of organizations are familiar, and therefore they make good examples. But there are many other conceptual spaces that locate and organize agent interactions. All that is required is that the concepts convey a sense of multiple categories that can be the "locations," that agents in the population can be members of different categories (and thus have different "locations"), and that the "locations" convey something about the likelihood that agents will interact. A social system of castes, or classes, or statuses can serve as a conceptual space. It seems poignantly clear that agents labeled "untouchables" may be restricted in their patterns of interacting even with those who are quite nearby in physical space. To give one more example, while nations may be thought of as regions of space-time in which agents are located, *nationalities* are conceptual categories. Israeli and Arab nationals living in New York City may systematically avoid each other even though they live only a few blocks apart.

We have stressed that interactions are located in both space and time. But we must reiterate an additional point about time: in Complex Adaptive Systems the sequential ordering of events can have huge effects. A change that increases proximity, that makes two agents more likely to interact, means that on average the interaction will occur sooner. If it takes place before events that it would otherwise have followed, it may change the character or likelihood of those events. The system can have an entirely different history as a result.

Example: Combating the AIDS Virus, Part 1

While the themes of proximity and activation may seem abstract, they can have important consequences. An excellent example is research on the spread of the Human Immunodeficiency Virus that causes AIDS. One of the great scourges of humankind, this virus infected over forty million people and killed nearly twelve million in the period from 1983 to 1997. As we write, there have been recent dramatic improvements in our understanding of how the virus acts and in treatment for those afflicted. Hopes are rising for prevention possibilities. The first tests of vaccines are about to begin. The story of these improvements involves many dedicated and inspired researchers in a wide range of fields. But the part of the story on which we want to focus is the new understanding of the spread of the disease developed by Carl Simon and his colleagues John Jacquez, James Koopman, and Ira Longini. (For a popular account, see Burr, 1998. For details see Jacquez et al., 1994, and Koopman et al., 1997.)

Their work began from the observation of a contradiction between what we knew about the spread of the virus and what was assumed about disease spread in classical epidemiological models. On the one hand, it was recognized early that HIV traveled disproportionately within groups such as intravenous drug users, homosexual males, and prostitutes. Social structures channel people in these groups into high rates of virus-transmitting interaction with some members of society and into quite low rates with others.

On the other hand, the best developed models of disease transmission assumed that the population of agents mixes randomly. These models make the unrealistic assumption that the next contact between two agents is equally likely to involve any two members of the population. Clearly, departures from this assumption could be important. Models must be used if we are to forecast how a disease will spread in the future. In order to make allocations of (limited) public health resources it is crucial to forecast which interventions will have the largest effects.

It may seem obvious from the perspective of harnessing complexity that there could be large consequences of more realistic assumptions about what we call "the interaction of types." However, assumptions of random mixing are well established in the disease-modeling community. Models based on those assumptions have provided many very important insights, and they have the advantage of being manageable mathematically. This is no small matter. A model with "better" assumptions that cannot be used to generate predictions may be of little value. Much work in the HIV-AIDS field went ahead relying on estimates that assumed random mixing. A major contribution of Carl Simon and his colleagues was to develop models of HIV spread that assumed nonrandom mixing and could still be fitted to data and made to yield useful predictions. In our terms, they found a way to build into the models more realistic patterns of interaction among agents.

Their work advanced the understanding of HIV infectiousness, the probability that when an infected individual meets an uninfected one, the virus will be transmitted. Using their assumption of nonrandom mixing to reanalyze available data, they concluded that infected individuals were much more infectious during the first few weeks after acquiring the virus than previously estimated—perhaps a thousand times more infectious! At a later stage in the infection, when their immune systems had begun to respond to the virus, they were estimated to be far less infectious than previously believed.

These new estimates had important public health implications, and they contradicted other models that were widely used. There was opposition—as often happens in research communities. But—as also often happens in science—agreement with other evidence eventually paid off. Among other things, more precise assays of the blood of infected persons came onto the scene. They showed astronomical counts of virus particles in days just after infection that dropped dramatically as the immune system began responding. The work is now hailed as an important advance. The paper in which Simon and his colleagues used a nonrandom mixing model to estimate transmission probabilities (Jacquez et al., 1994) won the 1995 Temin Award for Scientific Excellence in the Fight Against HIV/AIDS. Understanding the spread of HIV as a Complex Adaptive System is allowing officials to make new estimates of the effectiveness of public health interventions.

•   •   •

Now that we have laid out a more general view of proximity and activation, and have considered a rich example of how much they can matter, we are better prepared to discuss the mechanisms that can change interaction patterns. First, we will discuss the mechanisms that a designer or policy maker can use to change the interaction patterns of the agents. Then we will discuss the mechanisms that a single agent can use. Each kind of mechanism can be thought of as a form of filtering that selectively allows more interaction with certain agents and less with others. Each class of mechanism works on a different principle and achieves a different kind of screening of an agent's contacts. All the mechanisms move interaction patterns away from what the agent would experience if contacts were determined at random.

External Methods
of Changing Interaction Patterns

The principal mechanisms available to change interaction patterns from outside include the construction and operation of barriers to interaction—or removal of such barriers. We consider these mechanisms first, followed by methods for structuring the sequencing of activities over time.

Barriers to Movement in Time and Physical Space

We can see examples of barriers to interaction that have been erected all through our social world. These include city walls, national borders, prisons, monasteries, private clubs, computers that are deliberately isolated from outside networks, and middle schools that isolate those facing the onset of adolescence from others who are either younger or older.

The essential effect of a barrier to movement is to make some agents more proximate—more likely to interact with each other—and others less. Direct changes in physical space are the most obvious ways to accomplish this: walls and moats. But one can also alter the technology of moving through physical space. Not maintaining a road between two villages reduces the traffic between them. "Grounding" teenagers by taking away their car keys increases their interactions within the household and decreases their interactions beyond it.

All these devices have opposites, of course. For walls, moats, bad road maintenance, and grounding we can substitute doors, bridges, good maintenance, and car privileges.

Time can also be altered by controlling the technology for moving through it. Writing is a way of interacting with the future (as well as across space). Destroying written records is a way of depriving heretics of interactions with future agents. Reading is a technology of interacting with the past as well as with distant places. Preventing agents from learning to read can isolate them from thoughts developed in earlier generations (as well as thoughts from far away). Barriers in this sense can alter activation of agents by blocking interactions over time.

Many earlier information technology "advances" can be understood as reducing barriers to interactions across space and/or time. The technology of writing had this effect. In turn, it was greatly advanced by printing with movable type. This dramatically reduced the costs of the pattern of interaction in which the ideas of one source are communicated to many recipients across time periods and distances that were previously prohibitive. Other "broadcast" media, such as radio and television, have this one-to-many property as well. These technologies made major contributions to the formation of the large nation-states that dominated nineteenth- and twentieth-century world politics (Anderson, 1983). Especially when controlled by central authorities, they had enormous power to make diverse and dispersed populations more homogeneous in their knowledge, loyalties, and even language.

A major limitation of imposing and removing barriers and of other simple manipulations of proximity and activation is their imprecise selectivity. Crude physical boundaries rarely cluster together all the agents who would benefit and only those agents. Technological interventions that remove barriers often increase both wanted and unwanted interactions—as the World Wide Web brings us closer both to groups we admire *and* to groups we despise. Two further classes of mechanisms for modifying interaction patterns are extensions of the basic barrier approach that achieve greater selectivity. They involve barriers that are conceptual rather than physical and barriers that are "semi-permeable."

Barriers to Movement in Conceptual Spaces

Conceptual spaces are used by the agents themselves to make distinctions. Accompanying these distinctions there are usually boundaries. Movement in a conceptual space is not free of restriction. An agent in a hereditary caste system cannot simply pick up and move to another caste. Nor can an employee just move to a better-paying job at will. The barriers to movement are part of what defines the "location." Supervisory jobs that pay more have specific qualifications and may be subject to a competitive selection process. Castes are defined by socially maintained rules of entry and exit.

Earlier we concentrated on examples that involve altering processes of interaction across space or time. But conceptual barriers are among the most extraordinary human inventions for accomplishing similar goals. There are clan identities, club and fraternity membership criteria, citizenship rules, ethnic groupings, religious affiliations, and a host of other socially defined categories with hard boundaries.

These conceptual barriers place much more refined filters on patterns of interaction. Because they are conceptual rather than physical, their effects on interactions can be much more selective. Clan identity may dictate one kind of restriction on marital interactions (such as marrying within), another kind of restriction on commercial ones (such as borrowing money from outside the clan), and none at all on relations such as who can be employed. Moreover, the members of the clan may be dispersed in space. Although no physical boundary could effectively contain all the members and only the members, the concept of the clan can contain them and structure their interactions with each other and with other agent types. Interaction determined simply by coresidence within a common physical region cannot be so finely differentiated by types of agents. This is an enormous advantage for conceptual barriers as a means of shaping interaction patterns.

As always, there are disadvantages. Conceptual barriers, like their physical counterparts, can be causes of underexploration because they restrict interactions to homogeneous and familiar pools of other agents. Members of commercial firms often find that they have fallen into a pattern of talking about the business only with other members of the firm. The opportunities for learning about markets not being served are diminished when this happens. Arguments for diversity of religious and cultural types in public institutions often make the underexploration point, that there is much to be learned from interactions that are more heterogeneous.

Semi-permeable Barriers

A **semi-permeable barrier** is anything that prevents some kinds of interactions while permitting others. Often people can make an existing barrier or boundary selectively permeable. In fact, many walls have gates and guards who exercise selective control on comings and goings. Nations have immigration rules enforced at their borders; religions permit conversions on specific conditions; there are systems of e-mail filters that let in some messages and screen out the rest.

Indeed, the rise of the Internet has powerfully directed our attention to the design of semi-permeable barriers. There are such low costs to the movement of information in this medium that there is little indirect filtration of the kind accomplished by costs of movement in physical space. To move a letter to your house promptly, someone must pay first-class postage, which discourages first-class broadcasting of trivial communications. Even at low rates, postage charges on advertisers serve to filter out many messages, most of them unwanted. But, as we all now realize, plummeting costs of all forms of electronic transmission are contributing to monumental volumes of low-value or even harmful communications. The result is a boom in semi-permeable systems such as sophisticated network firewalls, V-chips for televisions, and the PICS metadata standard that facilitates advice and blocking in selection of Web sites (Resnick, 1997).

As the Information Revolution reduces the cost of moving information in "cyberspace," we lose a fundamental property of networks embedded in physical space. Those who are "near" you are not necessarily "near" each other. The classical character of conventional social structures is undone, as the mix of connections for each agent contains more relations that lead outside local, mutually connected clusters. This increase of out-group connections changes the spread of rumor and disease and reduces the correlation among the knowledge bases of interacting groups of agents. It also means that many who are near you, in the sense of being able to reach you easily, may not be like you, or be liked by you. The demand for selective permeability is great when an overwhelming number of interactions that consume time and resources are possible, and the other agents are not well known to you or to those you know well.

The introduction of sophisticated filters to achieve semi-permeable barriers in cyberspace is just the most recent episode in a long history of devices that add greater selectivity to simple barriers, whether those barriers are physical or conceptual. Our choice of the label "semi-permeable" reflects some of the earliest inventions of the biological realm, such as membranes that are able to admit some substances while screening out others. Not only do they function to increase proximity by establishing high concentrations of key resources inside the membrane; in some cases they also affect activation patterns by altering their selectivity as concentrations go above or below key thresholds.

Beyond such fundamental biological cases, there are many examples of semi-permeability in the social world. To the guards and immigration officers mentioned earlier we can add the secretaries who control access to their boss's calendar; the boards that test, accept, and expel members of professions such as law and dentistry; the automatic gates that block entry to a parking lot when it is estimated to be full; and the priests who enforce rules of religious "immigration."

In all these cases a barrier is conditionally opened or closed to an agent wishing to move through it. It is no accident that many of the examples involve delegating a person to make those choices. The selectivity is important and not easy to automate well—as anyone knows who has looked wistfully at an empty parking place lying just beyond a blocking automatic gate that "thinks" the lot is full.

The great advantage of semi-permeable barriers is the increased precision of blocking and permitting movement in physical or social spaces. Crude physical restrictions may amount to "no one may pass." Conceptual barriers can be more selective, providing more complex rules that allow desirable types to pass. Semi-permeable barriers can allow passage where admission should be governed by momentary conditions, such as parking lot fullness, or in situations where rules cannot cover well all the circumstances that may arise— as may happen at the door of a highly popular discotheque.

The great disadvantage is the possible mismatch between the rules or criteria governing the selective admission and the long-term welfare of the system behind the barrier. A religious community may have too few children born or surviving, but the priests may stick to strenuous tests for converts, driving membership below viable levels. The automatic gate may have an error-prone method of estimating the occupancy of the parking lot. Cell walls may admit virus particles with a shape that resembles a needed protein. Both conceptual and semi-permeable barriers may admit the wrong agents or block the right ones.

Example: Combating the AIDS Virus, Part 2

Earlier we recounted the enriched understanding of the spread of HIV that has come from appreciating the role of interaction patterns. There is more to the story, and its second half illustrates the idea of external interventions in patterns of interaction. Now that views on HIV infectiousness have changed to acknowledge its very high levels in the few weeks after it enters the body, important new questions have been brought to the foreground. A major one is how to evaluate the effectiveness of potential vaccines.

From the point of view of an individual receiving a vaccine, the important questions are whether it will lower the chances of acquiring the virus or will reduce the severity of the illness. But from the point of view of the population and its interactions—the viewpoint we are urging here—there is also the question of whether the vaccine will alter the probability that an infected person will pass the virus on to others who are uninfected. Acquiring the virus and transmitting it are different processes. It could happen that a vaccination might not fully protect those receiving it but could still be able to stop the spread of the virus through the population by diminishing infectiousness, the likelihood that the infected person will pass the virus on. The work on nonrandom mixing, by dramatically altering understanding of infectiousness, has now contributed to revised ideas about how to evaluate vaccine trials. As a result, it will be possible to choose among vaccines based on a better evaluation of how each treatment would affect the total number of people who will become infected. Through the improved ability to appraise those consequences, millions may be spared the suffering and death of AIDS.

Carl Simon is a veteran of analyzing Complex Adaptive Systems in our BACH group. Together with many others, he has translated ideas about complex systems into the most tangible results, the health of human beings.

Activation in Sequence or in Parallel

Most of the mechanisms discussed thus far have involved deliberate manipulations of spaces, either physical or conceptual. But a manager or policy maker sometimes has the opportunity to manipulate the role of time more directly. An example of this occurs in process design, when a complex new product is being developed. In such a situation there are many components of the overall design, and the configuration of each one depends to some degree on the designs of several of the others.

The conventional approach to such design interdependence has been to work in sequence. First have a team design, say, the frame of the car and its body shape, then perhaps the engine, suspension, the interior, and so on. Different sequences are possible. Part of the art of managing sequential design is to choose a sequence in which the discoveries of problems at the later stages do not force too many changes in the results of the earlier stages (Womack et al., 1990).

There are a number of drawbacks to this approach. A major one is that it requires a long elapsed time from the beginning of the first design stage to the completion of the last. In markets in which competition centers on the introduction of new models, there are important competitive advantages to shortening the design cycle. This is why automobile companies, for example, have moved increasingly toward a parallel rather than sequential process of design. In this approach the designers work on all the components of the car at the same time. The project is completed in the time required to design the most time-consuming single component.

This rearrangement may offer a large improvement, but only if the communication and coordination among all the simultaneous design subprojects does not slow them to a crawl or introduce inconsistencies. From our point of view, the gain occurs (if it does) by a resequencing of the interactions among the agents in the design teams. Designers of a component that would have been developed at any stage in a sequential plan would have interacted with those "upstream" by receiving finished plans and perhaps even prototype artifacts. Their work would be contemporary with few other component design teams.

In the parallel version, many more design teams are active simultaneously. Communication among the subprojects in the parallel version cannot be so much via finished plans and artifacts but has to be more in terms of projections and models of what the components will be like in the end.

Computer-aided design tools help enormously to conquer the difficult problems of communication and coordination in the parallel version. In the most modern implementations of parallel design, a computer simulation is built of the entire product, say a car or an airplane. Each new version of a component is reflected in a change in the simulation, which must continue to show that the overall product still "works." The shared model is an extremely rich artifact that helps achieve the needed coordination.

The idea of converting processes that normally are serial into parallel ones is quite general and is seen in many parts of computer science as well as in industrial settings. Nature, however, got there first. A population of breeding animals is also a parallel design process, in which many designs (combinations of genes) are tried in a single generation, and parts of the most successful designs are passed on to the succeeding populations.

Internal Methods of Changing Interaction Patterns

We turn now to the methods of changing interaction patterns that are available to a single agent.

Following Another Agent

One of the simplest mechanisms that can modify interaction patterns arises from one agent's staying near another. The most basic examples of this mechanism involve staying nearby in a physical space.

As we shall see, however, the general character of the mechanism persists even when the proximity is conceptual rather than physical.

The biological prototype of this mechanism is adhesion, in which one organism sticks to another or stays close to it. It is seen all over the biological world, from a virus that sticks to cell surfaces, to a flea that visits the human world in the company of a rat, to a baby kangaroo that travels with its mother by staying in her pouch. The effect is that the "following" agent experiences a pattern of interactions similar to that of the "leading" agent. In addition, there is also more interaction between the follower and the leader. In daily life we spend time with our relatives, co-workers, and friends, and by "sticking with them," we also meet the people they know.

There are many follow-the-leader mechanisms beyond these simplest ones. For example, there is apprenticeship, in which the apprentice stays close to, and shares many experiences of, the master of some trade. Beyond formal apprenticeship, there are still other forms of what has been called "legitimate peripheral participation" (Lave and Wenger, 1991). These arrangements not only let the trainee see how an expert individual works but also allow access to social interactions that are essential to the effectiveness of the leading agent. Other instances of modifying interaction by staying close to another agent include: hospital rounds; "big brother" relationships—either with real siblings or deliberately arranged mentors; following a guide around a tourist site or other new place; research training; going to work with a parent; or attending the school of a widely known teacher who has attracted other students with the same interests. All of these familiar procedures of the social world, and many more, share an element of acquiring the interaction patterns as well as the strategies of a leading agent, who serves as a kind of template.

In the world of computer networks, this kind of mechanism has been generalized. "Recommender" systems allow users to "adhere" to the tastes of others, in order to interact with the persons and objects they have encountered. In such systems, the user provides some profile of interests, say by rating a sample of films. Then the system tells the user about films that were liked by other raters whose pattern of evaluations is similar to the user's own. Comparable methods have been constructed for finding other "taste goods," such as books and music, for finding professional assistance (dentists, stockbrokers), and for finding on-line discussion groups or World Wide Web pages of interest (Hill et al., 1995; Resnick et al., 1994; Shardanand and Maes, 1995). These electronic versions imitate the wisdom of the now faded time when library books had signed checkout cards and it was possible to see who had previously read a book. In the contemporary on-line versions, however, you may not need to recognize the names of the others. Indeed, the Information Revolution makes possible recommendations based on a statistical synthesis of others that might be closer to predicting your tastes than any other single user, or even a professional critic. Such systems are often able to help users find other agents or objects they will enjoy (Gladwell, 1999).

These mechanisms for following an agent present an intricate mix of advantages and disadvantages. Among the sources of benefits and problems, we focus on two. The first is the ability to acquire interaction patterns implicitly without having a good theory of how things work. The second is living in the kind of clustered social network that results from wide use of the mechanism, a network where many of the other agents have strongly overlapping knowledge and social contacts.

Using mechanisms for following, an agent can tacitly pick up the contact pattern of a leading agent without necessarily understanding the causes or the effects of that pattern. Although there are problems that we return to below, not having to understand the situation can be an important advantage. Indeed, most of the accomplishments of biological evolution, and much human social change, have occurred without the benefit of such explicit knowledge, let alone theoretical understanding. Nature can make a quite efficient food web without the science of ecology. Of course, theories are powerful when we can achieve them. (With scientific understanding, we could have foreseen the consequences of actions like introducing rabbits to Australia, where natural predators were absent.) But good theories are extraordinarily costly to create and share with others. For many complex domains, they may long remain beyond our capabilities.

In social systems, most transmission of traditional knowledge uses the approach of learning how things are done—or ought to be done—without understanding fully the reasons why. Work practices, trading partners, religious ceremonies, musical forms, and social role obligations, are all passed along in this way—to take but a few examples. For the most part, knowledge transmitted in this way serves people well, even if it may carry along some counterproductive beliefs. The mechanism of copying the interaction patterns of other agents passes along vital social knowledge and allows an agent to adapt, without requiring an explicit understanding of very complex social systems.

As with biological evolution, problems can arise when interaction patterns are transferred to new contexts, since the selectivity of a more precise theory is not available to sort out which features should be modified and which retained. Seasonal festivals that were highly functional can end up being celebrated at inappropriate dates because a religious calendar is not synchronized to the local climate, as happens with the planting holidays of the Northern Hemisphere religions that are now practiced south of the equator. Structures of family obligation that evolved in a long era of agricultural work and low spatial mobility can work badly when transferred to the highly mobile urban life of recent decades.

Copying another agent has a further important effect, in addition to picking up the other's pattern of interactions. At the population level, copying others' interaction patterns also introduces strong correlation among the contact patterns of the agents. If most agents are building their interaction patterns by such mechanisms, the resulting social system will have the cliquish property that most of those interacting with a given agent will also interact with each other. When social structures arise among agents situated in physical space with high costs of travel, this is the expected result.

Many advantages result from the formation of a social network with these correlated properties. Agents in such a population will have a large overlap with the contact patterns, and therefore with the strategies and knowledge, of most of the agents with which they interact. This overlap implies shared assumptions and common understandings, and these in turn simplify transactions of all kinds. Explanations can be brief and suffer few misunderstandings. Consequences of actions can be more correctly anticipated. The ease of communication in overlapping networks can help build social capital.

Agents' reputations for trustworthiness will be based on many previous interactions with many other (nearby) agents. If an agent behaves badly, it incurs heavy costs in loss of reputation since its contacts can be easily informed. This is another way in which clustered networks build social capital.

There are also disadvantages to the social structure that accumulate through pervasive agent-following. It can result in loss of informational diversity. In populations where any agents' friends, relatives, and co-workers also know each other, there can be loss of variety in the information easily available to a member of the group.

Mechanisms of establishing interactions such as we have discussed here, that work by taking other agents' patterns as templates, will tend to build social networks that are strongly clustered. That can have the side effect of reducing an agent's ability to explore a wide space of options. The result, if only these mechanisms are active, may be insufficient exploration and danger of premature convergence.

Empirical research shows that surprising or important information, such as news about job opportunities, usually does not come from people who are part of your closest group of friends. Instead, it arrives from "acquaintances" who are on the edge of your social world (Granovetter, 1973). Frequent interactions among friends who all know one another apparently leads to reduced diversity of the information they hold as a group. Your current friends would tell you about a perfect job for you in a distant town, but they are unaware of it for the same reasons that you are. However, your old college friend, who lives there, can be a source of significant informational variety.

As the data on job-finding show, a healthy social network should probably contain a mix of strongly and weakly clustered contacts. That is what would provide agents with a better balance of exploitation and exploration. And just such a mix is what is actually found in investigations of social networks. It is sometimes known as the "small-world property." It can be shown that a modest proportion of ties to distant others suffices to "shrink the social world" dramatically (Watts and Strogatz, 1998). While individuals who are far apart in physical space or social class seem very unlikely to interact, in fact they are usually separated only by a short chain of social contacts. The research of Stanley Milgram in the 1960s first clarified this closeness. It has entered the language with the phrase "six degrees of separation" as a shorthand for the idea that a short chain of contacts (perhaps no more than six) will generally suffice to connect any two people in the world (Milgram, 1967; Gladwell, 1999).

An important question is now coming into focus as the Information Revolution penetrates our societies: Does this "distance independent" technology change the mix of clustering in our social networks, giving us more contact with distant persons who do not know the others that we know? The underlying issue is not new, of course. At least as far back as the Roman roads, each gain in information transmission has reinforced the shrinking of the social world. The reduction of effective distance has continued with postal systems, telegraphs, and telephones. But the concern remains that there could be a "threshold effect," a level of connection among "distant" persons at which suddenly the total diversity of the *world* begins to decrease rapidly, even as the diversity impinging on individuals may continue to increase. Some fear that such a threshold might already have been crossed, even before the Internet has had its full effect. Among the data they cite as evidence are the booming worldwide markets for Euro-American cultural materials such as music, film, and clothing, which grow at the expense of local traditions, and the mounting rate of extinction among the world's languages (Doyle, 1998).

It may seem a paradox that as individual agents experience more diverse contacts, the system can become less diverse. But there need be no contradiction. As interaction patterns in a social system become less clustered, giving individuals the experience of interacting with more "distant" others, information (or rumor, or disease) moves more rapidly throughout the entire system. The world becomes so "small" that the sociological "islands" vanish, unable to keep their local ways and remain untouched by events elsewhere. Research has shown that—at least for biological populations—this is not the most favorable situation for adaptation. Early innovations spread too fast, and variety that can provide later improvements is lost—the phenomenon we have called premature convergence. The ideal breeding ground for novel life-forms seems to be an archipelago or a network of mountain valleys. In these settings semi-isolated populations breed with relatively infrequent exchanges of animals. Improvements occur but spread slowly enough to avoid a too-rapid loss of diversity (MacArthur and Wilson, 1967; Tanese, 1989).

Of course, we cannot choose actions based on a simple analogy of human social systems to breeding biological populations. The mechanisms of reproduction are very different, as are the criteria for assessing change. Nevertheless, lowered clustering of social networks may well increase homogenizing pressures (Axelrod, 1997). Comparing the costs of lost diversity with other effects is not our subject here. Deciding whether to resist or facilitate the loss of world social variety requires assessing the impacts on such dimensions as chances of nuclear war, spread of disease, efficiency of global business transactions, loss of cultural variety, or possibilities for concerted world initiatives on environmental issues. Our aim here is to pose the question in terms of the structure of human interactions, so that costs, benefits, and interventions can be thought through more fruitfully as issues arising in a Complex Adaptive System.

Following a Signal

A second strategy that an agent can use to alter its interaction pattern is to follow some detectable signal, moving toward locations that have better value. As in the case of following an agent, the movement creates new patterns of interaction for the following agents. From the point of view of an external designer, this strategy offers the opportunity to create or modify the signals that agents follow in order to alter their patterns of interaction.

People and other agents move in space toward desirable signals. They seek homes with clear air and low noise, near well-maintained schools. They frequent restaurants that are popular and busy and view movies that are highly rated. They seek jobs with higher pay. Some household robots have been programmed to monitor their batteries and head for electrical outlets when their power is low.

All these are patterns of moving through spaces (physical or conceptual) by following a signal. They have the direct effect of bringing us into a situation that is more desirable, but also the indirect effect of bringing us into the interaction pattern prevailing in the new location. We live in a more attractive neighborhood, and we are surrounded by new people who also find that neighborhood attractive and have been able to take up residence there. We enter their social networks. Occasionally, this latter effect is the predominant one, as in the crowded restaurant example. Artists move to one of the city's run-down quarters because that is where other artists are densely congregated around inexpensive warehouse space, a recurring urban process that is nicely documented in Stewart Brand's *How Buildings Learn* (1994). But more commonly, we move along a gradient for its own sake, to get a quieter apartment or to make more money. Then we experience the indirect effects on our interaction patterns, which may have been only dimly anticipated.

A disadvantage of following a signal is that an agent can get stuck on a local maximum, and not find the global maximum. If the signal can be detected only in the immediate neighborhood where the agent is, this can be a real problem. To compensate for it, the agent might have to make some large exploratory moves to determine whether the signal should be picked up and followed again from some different starting place. You can readily find the best bench in the park, but there could be a better one in some other park.

Getting stuck on a local maximum is a common problem in Complex Adaptive Systems. It can happen in abstract spaces as well as physical ones. For example, during the 1980s, rival groupings of computer companies formed to advocate different standards for the Unix operating system. They seem to have developed their coalitions by a logic of considering only small changes in the space of possible coalitions (Axelrod et al., 1995).

Now that we have some examples of movement by signal following, we can consider why it is different from the previous case, movement by agent following. Signals are usually associated with locations rather than agents. Restaurants are crowded. Neighborhoods are well kept. Jobs have pay scales. This is an important contrast with agent following. If you make friends with someone who is a good musician, you may meet people who visit the restaurants musicians like. However, your friend may choose places on a completely different basis, unrelated to a signal such as whether they are popular and busy. If your friend likes jazz, you may meet jazz lovers who go to restaurants where it is played. You are less likely to meet people who like whatever restaurant is currently in vogue.

Following a signal relies implicitly—and sometimes even explicitly—on a belief that the signal goes together with consequences an agent will prefer. Some of those consequences are direct—lower noise, to return to the example of choosing apartments. But usually there are indirect consequences as well, and among those are interactions with others who follow the same signal. In a building with quiet apartments you may have neighbors who prize quiet. At the tops of treacherous mountains, you meet people who are devoted to mountain climbing.

Signal following leads to locations that attract others who follow the same or related signals. Agent following leads to others an agent interacts with, through whatever mechanism. The former leads to interactions associated with the signal, the latter to interactions associated with the leading agent.

Often agents themselves have a property that can serve as a signal to other agents. Amazingly enough, even if such a property is assigned to the agents completely arbitrarily, it can still serve to help the agents organize their interaction patterns and thereby achieve greater success (Holland, 1995). We call such a property a **tag**—an initially arbitrary property of an agent (say, a number between 0 and 1) that is detectable by other agents and that can be copied by other agents. Examples of tags might include accents and styles of clothing. Tags work by allowing agents to interact with others having desirable tags—in many cases this will be tags similar to their own. When agents with mutually desirable tags interact a lot, they create a neighborhood. Even if those in a neighborhood initially have nothing in common but their tags, the fact that they interact with each other more than with others can make a big difference. We turn now to an extended example of how tags can establish neighborhoods.

Example: Tags in the Prisoner's Dilemma

Our colleague Rick Riolo has provided a beautiful set of simulation experiments showing how tags can function. The experiments demonstrate the power of following a signal in a conceptual space (Riolo, 1997). They show that arbitrary tags assigned to adaptive agents can become meaningful and serve to organize their behavior. In a computer simulation, he created simple agents that met each other and played the well-known Prisoner's Dilemma game. This game has a devilish structure. At each moment of action, each agent may be either selfish or cooperative. In the short run, it is always better for an agent to act selfishly, no matter what the other does. But if both agents can sustain cooperative behavior over a longer period, the cumulative result is better for both of them than a long run of mutual selfishness. The agents in Riolo's simulation were adaptive. They learned strategies for playing the game by emulating the strategies of other agents they met who had been successful. In effect, the previous success of agents they encountered was a kind of signal, which the agents could observe and follow.

If agents in a population meet each other at random, the usual outcome in such a simulation is that there is almost no cooperative behavior. The reason is that initially the agent interaction patterns are disorderly; each agent's environment consists of other players whose strategies were selected at random. In such an environment, those who happen to follow the rule of not cooperating at all will get the highest score. Conversely, those who experiment with cooperative actions do badly because they meet many unresponsive agents. With these coevolutionary dynamics, selfish strategies spread through the population, becoming endemic.

Riolo experimented by adding one feature to this well-known system. He associated a tag with each agent (actually just a random number between zero and one), and the agents were programmed so that they tended to avoid playing the game with any agent whose tag was not fairly close to their own. Thus an agent's strategy had two parts: a *tag* that helped determine who it would interact with, and a *rule* for making choices within an interaction. At first, the introduction of tags contributed nothing, since the tags were assigned with no relation to the selfish or cooperative tendencies of the agents' rules. Among the randomly assigned rules, some agents used "Tit for Tat." This is a rule whereby an agent cooperates at the beginning of an interaction, and then answers cooperation with more cooperation, and selfishness with more selfishness (Axelrod, 1984). When two Tit for Tat players had similar tags by chance and encountered each other, they both did very well. As a result, their strategies, including both their tags and their rules, were imitated by other agents who met them.

Tags are locations in a conceptual space. Imitating another agent's tag had the effect of moving a player to its location in "tag space." As agents moved to the locations of the successful Tit for Tat agents and adopted their rules, the neighborhood in tag space began to fill with agents who cooperated on the basis of reciprocity. These agents used their increased proximity in tag space to interact heavily (and profitably) with each other. By forming cooperative neighborhoods in tag space, they achieved levels of mutually beneficial cooperation that were impossible when contacts were random. Indeed, the length of the interactions had been chosen to be short enough that strictly random interactions would not evolve cooperation. Nevertheless, with tags, these short interactions were sufficient to help cooperation emerge. Initially arbitrary tags had taken on a meaning within the simulation, allowing agents to selectively play with others who tended to reciprocate cooperation.

If through chance variations an occasional purely selfish agent entered such a neighborhood, it did not get the benefits from repeated mutual cooperation that the Tit for Tats got with each other. Therefore, the rare selfish agent did not do well enough to be imitated. In fact, in such a neighborhood, selfish agents wound up "converting" to the Tit for Tat rule.

Although the tags Riolo used were numbers, in the social world any observable feature of agents can serve the same function: allowing agents to sort themselves into pools where they interact with compatible others and avoid incompatible others. Clothing styles, spoken expressions, bumper stickers, haircuts, doorway ornaments, and a host of other visible symbols have served to "break the symmetry" of interaction, allowing the compatible to achieve greater proximity. The interesting thing is that tags allowed compatibility of rules to be created from within the population itself. The key is that imitation of the tags of successful others helped neighborhoods to form, focusing the interaction patterns of an agent among those with similar tags, making the tag a meaningful basis for selective cooperation.

Unfortunately, the agents in Riolo's simulated world did not live cooperatively ever after. Riolo had another piece of realism in his simulation. Agents made errors in copying strategies, and so over time rules accumulated in a cooperative neighborhood that were purely cooperative instead of Tit for Tat. These agents behaved cooperatively no matter what another agent did. If the other agents were Tit for Tat, this did not make any difference; all agents realized the benefits of cooperation. But once a substantial fraction of the neighborhood was occupied by purely cooperative agents, the chance entry of a purely selfish type triggered a very different history. Instead of encountering Tit for Tats, doing badly, and being converted, a selfish player now encountered several pure cooperators, which it exploited for a very high score. It did so well that it was imitated, and some of its "converts" also encountered pure cooperators and did very well, causing more agents in the neighborhood to imitate the purely selfish rule. An avalanche of selfishness imitation occurred, wiping out the once cooperative neighborhood.

The entire system then limped along with very little cooperative action until by chance a few Tit for Tats again carried similar tags. Then the story recycled itself at this new location in tag space. The small cluster of agents who used Tit for Tat in this new location drew imitators to their neighborhood, which prospered until copying errors pushed the density of purely cooperative rules too high. Then the ability of Tit for Tats to "police" the new neighborhood against the purely selfish rule was undermined. The new neighborhood became vulnerable to invading purely selfish rules. And the cycle ended with another collapse into widespread selfishness.

Riolo was quite careful in interpreting his work. He did not claim that the introduction of arbitrary tags would always allow populations to form cooperative neighborhoods. Nor did he claim that all such neighborhoods are destined to break down over time. He showed that these dynamics did occur, and he documented the conditions that contributed to them. But he also showed that under other conditions they would not occur. For example, rates of copying error can prevent cooperation from emerging by being either too high or too low. This illustrates a characteristic virtue, as well as a difficulty, of simulating a Complex Adaptive System: the results can often be shown to depend on parameters like the error rate—just as they do in the real world.

While Riolo's results do not tell us what will always happen, they do show us one way that complexity can sometimes be harnessed to help a population break out of mutually reinforcing selfishness. He also shows us a weakness of tags as a method of harnessing complexity: they are not able to maintain "policing" of the cooperative neighborhoods they create. Since tags can be imitated, any agent could enter a successful neighborhood, by copying the tag of a member—even an agent with a selfish rule for playing the game. In the early history of a neighborhood, its many Tit for Tats converted such an entrant. But with the passage of time there were more pure cooperators in the neighborhood, and then an entering agent with a selfish rule unleashed havoc. The very copying errors that stimulated the emergence of cooperative neighborhoods later led to their dissolution.

The success of this tag space method of harnessing complexity depends on mutual support among the rules of the agents. The arrival in the neighborhood of more Tit for Tatters benefited those already there. A similar dynamic can occur in other cases. In the infancy of a social movement, for example, each new supporter who is gained increases the value of the movement for existing supporters and increases the attractiveness of the movement to those who might subsequently join. However, Riolo's example also shows that things will not remain simple if some rules exploit others. Then a signal can be followed to build a concentration of one kind of rule, but it may subsequently serve as a guide to agents who would prey upon that concentration. Senior citizens may name an Internet newsgroup for discussing their common interest in travel. But later, when they have built up its membership, unscrupulous fraud artists can use the list to sell worthless vacation bargains.

•   •   •

Beyond the work on tags that we have described, there are many other dynamics of neighborhoods based on signals. For example, cases occur among fashion leaders and fashion imitators in many fields such as clothing, music, hairstyles, and advertising techniques. In this interesting form of signal following, the logic is that some agents want to be among the distinctive few. Eventually, however, the fashion leaders attract others who want to be like them. This imitation reduces their distinctiveness. Finally, the fashion leaders must move on to new bases of distinction, only to be followed again.

As with the mechanism of following a leading agent, the mechanism of following a signal has both advantages and disadvantages. On the positive side of the ledger, the followed signal itself sets a context for interpreting what happens. If you worked to obtain a promotion and now you are enjoying your new colleagues, the improved job status may seem a likely cause. The signal can be useful in contrast with the agent-following case above, where the context determining the interaction patterns may not be established so explicitly. As a result, signal following favors an approach that we will describe in the next chapter as the attribution of credit. The consequences of the resulting patterns of interaction can be evaluated in terms of criteria that are more specific. One can say, "I wanted a restaurant that would be lively, and it was." In an agent-following strategy, it is more difficult to know what criteria to apply to the resulting interactions. If you go back to the restaurant that you visited with your friend, you may have a sense that you liked it, but you may not be sure exactly why.

A weakness of signal following is that the signal may be a bad predictor of the quality of the interactions that follow. The residents of a more expensive neighborhood may prefer privacy or travel heavily and so they don't interact with you. The signal may not be causally associated with the interactions experienced, as when your child goes to a better school but has a bad year because of a teacher's undiagnosed health problems. Since you do not know the true cause, you might conclude that the problem stemmed from some distinctive feature of the school you thought was "better," such as its (oversized?) new building or its students with (overly?) ambitious parents.

The two mechanisms of following an agent and following a signal do not exhaust the possibilities for changing interaction patterns by moving through a space. They are, nevertheless, the most important cases. To see another approach to changing interactions, we turn now to mechanisms that involve changing patterns of activation. The interactions among agents in a Complex Adaptive System can be dramatically altered simply as the consequence of changing the likelihood of agent activity.

Forming Boundaries

Consider first how internal activation can form patterns such as the patches and stripes of animal hides. Examining this very simple case will reveal a striking general principle. How do pigments end up concentrated in one area of skin and absent nearby? What controls the size of the patches so that a Holstein cow looks so different from a Dalmatian dog? And why does neither form stripes like a zebra?

There is a long history of work on how these patterns are created. An overarching insight extracted by researchers goes under the acronym LALI, for "local activation, long-range inhibition." In the vast majority of cases of patterns on animal hides, a pigment (or other pattern substance) that has been deposited in an area makes it more likely that another, similar deposit will occur nearby, and less likely that a similar deposit will occur farther away (Bonabeau, 1997). The details of how these two influences spread can account for the sizes and shapes of patches, but the basic principle holds across many different patterns.

The combined effect of local activation and long-range inhibition is especially powerful. Nearby areas will have many pigment deposits and, in consequence, areas farther away will have strong pressures to be clear. Together these two pressures lead to the formation of sharp boundaries for the patches.

The principle is not confined to the skins of animals. After all, they are just one kind of space. Local activation together with long-range inhibition will work to establish patchy ethnic neighborhoods in residential areas of cities, or splinter groups on a political left wing, as well as spots on Dalmatians' backs. It is plausible to suppose that members of an ethnic group may like to live near others with similar cultural backgrounds. So a local cluster of ethnically similar individuals will be attractive, leading to more individuals moving in nearby, which will enlarge the cluster. This corresponds to local activation. It is also easy to suppose that neighboring areas may become resistant as they perceive the rapid growth of a group whose culture they do not share. If the neighboring area makes it harder for members of the ethnic group to move in, this will act like long-range inhibition. The result will be the kind of sharp boundaries of ethnic neighborhoods that are so striking in cities like New York. The story can be told again, with the underlying space being neither a skin nor a city but a political spectrum with concentrations of ideologically like-minded individuals who attract those who share their views and oppose the "heresies" of those who do not.

Separating Time Scales

Our examples have centered largely on movement (and barriers) in spaces, physical and conceptual—with the exception of our short discussion of reading and writing as technologies of interaction across time. A quite different mechanism has been pointed out by Nobel laureate Herbert Simon, in his classic essay "The Architecture of Complexity" (1981). He examines the tendencies of many biological and social systems to assume hierarchical (or "pyramidal" or "inverted tree") shape. Simon notices that the upper layers of such systems typically involve processes that span longer time intervals, while the lower levels are more often involved with processes that run relatively quickly. CEOs and board members concern themselves with the question of what markets should be entered in the coming years, while factory floor supervisors concern themselves with the production schedule for the coming week. A brain may take one or two seconds to compose a sentence, but the nerve cells within one of its many cooperating parts, say the left cerebral cortex, discharge in times measured in milliseconds.

Simon argues that this hierarchical arrangement of time scales supports effective governance in a system, which is why it is often seen in armies and bureaucracies. The reason is that the slower activity at the upper levels establishes a stable context for faster processes running at lower levels. It helps in providing a social service if the definition of who the client is does not change while you are providing it. Likewise, it helps in taking a defended hilltop if the definition of the enemy does not change while you are attacking it. Hierarchies have the property that every element of the system (but the top one) has just one supervisor, and most elements (but the bottom ones) have several subordinates. So whenever a "superordinate" element acts, it establishes a context that allows its subordinates to act in concert. This is tremendously useful in achieving the benefits of coordination. Napoleon had the same idea when he said, "One bad general is better than two good generals."

In Simon's view, this separation of time scales is so advantageous that we should expect evolution to produce it frequently. Actions with long time frames should tend to become assigned to positions that govern levels in which actions have shorter time frames. Systems that organize this way will have a competitive advantage, and there should be more of them over time (Simon, 1981).

Simon stands far back to discern this tendency across a wide range of what we call Complex Adaptive Systems. But standing close-up to organizational cases shows that the assignment of actions to levels is generally done by agents within them. If those agents understand the principle, it offers them another opportunity to influence future events. One might argue that "in our consumer products business, products now should come and go very rapidly. The CEO should no longer make those decisions for our company. At that level, the concern should be for our long-run reputation with consumers and for the research and development that generates new products. Our reputation should shape our products more than the other way round." As usual, we do not claim that an argument like this is always right. We do think that it comes from—and leads to— the right kinds of questions.

Redistributing Stress

We turn now to a mechanism that depends on interactions within the system to stimulate further activation of agents. A good starting point is work by Per Bak and his many colleagues on "self-organized criticality" (Bak, 1996). They have studied a wide range of systems in which some kind of stress propagates through the system. Examples include sand piles and snow fields (which release in avalanches), and underground rock layers (earthquakes). In these cases, the systems consist only of artifacts. Later work has shown that some of Bak's results can apply to systems that include agents—for example, people crowding their cars onto highways (traffic jams), or ecologies of species (mass extinction events). What is happening in all these instances is that a small event may or may not trigger other events. An additional grain of sand added at the top of a pile may dislodge another, which may dislodge others. The added particle may dislodge few others or none. When sand is piled at its "angle of repose," it is ripe for avalanches. Bak and his co-workers have investigated the sizes of the avalanches (or earthquakes, or other stress releases) observed in such systems. They have repeatedly found a striking pattern, known as a power law distribution, in which the numbers of events in different size categories are related by a constant proportion.

This principle can be illustrated by the distribution of the sizes of wars as well as the sizes of avalanches. For example, there are many small wars, a moderate number of medium-sized wars, and a few very large wars. This pattern of sizes of wars can be seen as the result of propagation of stress resulting in a power law distribution. Consider these pioneering statistics (Richardson, 1960) on the distribution of wars from 1820 to 1945 that caused 300 or more battle deaths:

Size of War	Number of Wars
About 1,000 deaths	188
About 10,000 deaths	63
About 100,000 deaths	24
About 1,000,000 deaths	5
About 10,000,000 deaths	2

Note that for each tenfold increase in the magnitude of the war, there is roughly a threefold decrease in the number of wars. However, the increase by a factor of ten in the magnitude means that the wars in the category kill a total of about three times as many people.

There are other important results about these systems. One is that after they build up to their critical state, such as the sand pile's angle of repose, a long time without a big event does not necessarily mean that something big is due soon. There is such a complicated interdependence among all the sand grains, snowflakes, or species, that you cannot know whether small events are relieving or increasing the stress. Bak offers a very interesting observation about the consequences of the accumulation of interdependence: "In the critical state, the sand pile is the functional unit, not the grain of sand" (Bak, 1996, p. 60; see also Schroeder, 1991).

If sand particles, or species, or highway drivers keep arriving in the system, they will soon build it back up to its critical state, even if there have been big releases. Bak therefore introduces the idea of self-organized criticality, the tendency of the system to stay near its critical state.

Example: Modes of Failure in Information Systems

To understand a system that relies on highly interactive elements, whether agents or artifacts, one needs to take into account the possibility of major stresses that could lead to large-scale failures. The possibility of large-scale failures is an especially important problem for the information domain for three reasons. First, information systems have typically undergone very rapid evolution. They often involve both new technology and new institutional arrangements. Therefore, they have not existed long enough for the development of a good empirical foundation for risk assessment and management. Second, information systems continually undergo major changes, so that a good empirical foundation for risk assessment and management may never become available. The final reason that the possibility of large-scale failures is important in information systems is that vital economic and military functions are highly dependent on these systems.

As we shall see, some types of failure are well understood and relatively easy to design against. Other types of failure are less well understood and more difficult to design against.

The simplest type of challenge occurs when failures are strictly independent. For example, if a basement room is lit by four lightbulbs, and one burns out, the others will continue to supply light. The loss of one bulb does not make the others fail any sooner than they otherwise would have.

The primary method of risk management for independent failures is to build redundancy into the system. In the case of lightbulbs, enough bulbs are used so that if one (or even two or three) burn out, there will be enough light for activities to continue until the bulbs can be replaced. In a more sophisticated manner, redundancy makes possible reliable traffic flows through information networks by channeling traffic around nodes that fail. In addition to redundancy, a useful design feature to deal with local failures is to avoid having any one element of the system be essential to its overall performance. This is typically achieved by making the system highly decentralized, like the Internet. Indeed, one of the primary advantages of Complex Adaptive Systems over more rigidly centralized organizations is their resilience in the face of local failures.

Design problems become much more difficult when local failures are not independent of each other. The first problem to consider is correlated shocks. These are failures that occur when the elements of the system tend to fail at the same time for the same reason. For example, suppose there are ten radio transmitters, each of which works 99 percent of the time. If the failures were not correlated, at any given time at least one of the transmitters would almost certainly be working. However, if sunspots are the reason the radio transmissions fail 1 percent of the time, then all ten radio links may well fail at the same time.

Typically, the design against correlated failures involves

identifying the sources of shocks that are external to the elements and that might therefore cause failures in several elements at the same time. Sunspots are such a source of shock for the radio transmissions. Once these correlated sources of error are identified, redundancy can often be attained by building new elements of the system that are not susceptible to these particular shocks. For example, landlines are not affected by sunspots. Even when the sources of external shocks cannot all be identified, a general principle is that the more diverse the elements, the less the chance that they will all be vulnerable to the same kinds of external shock. Diversity in Complex Adaptive Systems not only allows exploration of new options but also provides resiliency against common shocks.

The risk of monoculture, which we mentioned in discussing variation, provides a good example of the need to avoid common shocks. If vast tracts of agricultural land are planted with the same strain of a crop, then an unusual environmental condition, such as a new pest, can cause devastation. While monoculture may be efficient in the short run because it exploits the very best strain of crop, it tends to be fragile. Likewise, information systems that rely on widespread use of common hardware or software components also risk fragility. They provide vast, fertile targets for viruses and other virtual pests.

It is ironic that in our efforts to stabilize systems against independent or correlated failures, we often transform them into more tightly coupled systems that redistribute stress. For example, we create power grids so that regions can borrow power from neighboring regions. Local power shortages are reduced, but larger failures become possible, such as the cascade of power outages that caused the 1977 New York City blackout, or the two outages of 1996 that each affected millions of utility customers in the western United States.

Independent failures and correlated failures can both occur

in systems whether or not the elements are connected to each other. The third type, stress propagation failure, becomes possible when the elements interact naturally, or are designed to interact. Here the risk is that a failure in one element can cause stress in another element, leading to failure of that element as well. Eventually a cascade of failures could cause a large-scale failure. As we have seen in the section on redistributing stress, stress propagation failures occur not only in information systems but also in many other systems that are closely coupled. In fact, advances in information systems allow more and more systems of different kinds to be designed in ways that provide efficiency through a close coupling of their elements, with attendant risks of large-scale failures (Perrow, 1984). A good example is "just in time" inventory systems, which increase efficiency by reducing inventory buffers, but which also mean that a strike in a single plant can rapidly close a whole network of plants.

Unless the coupled structure of the situation is changed, interventions to stave off catastrophic releases can only be expected to be briefly effective. Snow fences, emergency interventions for threatened species, and efforts to control individual bad drivers on a freeway all avoid trouble only in the short run. For systems in the critical state, an event from some quarter will eventually trigger a huge chain of effects. Of course, one might be concerned mainly about what happens during the period when the preventive measure postpones a big release. The treatment of self-organized criticality does not argue that there is no postponement, only that a local intervention will provide no relief in the long run.

To change the basic character of the system, short-term interventions are not effective. The relative frequency of big and small events stems from the nature of the interdependence between the elements: the stickiness of the sand or snow, the variety of other species that a given species consumes, the reaction times of freeway drivers, or the borrowing privileges of power grid regions. These linkages among the artifacts or agents are the means by which events change the probability of future events.

While the design principles for systems that propagate stress are not well developed, several ideas do seem relevant. First, the entire problem can be avoided if the elements of the system can be prevented from transferring stress to each other. For example, if unmet loads from failing elements were not automatically passed along to other elements, cascades of failure would be prevented. Another, related, approach is to prevent large "avalanches" by partitioning the system and preventing load transfers from elements in one part to elements in another. A third approach is to build more slack into the system so that individual elements fail less often, making cascades of a given magnitude less frequent. All these methods work at some cost in lost opportunities for load sharing or other efficiencies. However, as we saw in our data on wars, rare large events can have extremely severe consequences. For this reason, it pays to search for effective ways to reduce stress propagation at the cost of only modest reductions of efficiency.

So far, our consideration of modes of failure in information systems has focused on "natural" shocks, whether they are local, correlated, or caused by the propagation of stress. We next consider shocks that are deliberately caused by attack from other agents in the system.

The most dangerous attacks are often ones that exploit some vulnerability in a surprising new way. For the attacker, surprise is frequently possible only by risking the revelation of the means of surprise. For example, using a new way of overloading a computer system might work the first time but probably not a month later. Thus, anyone who has the means to surprise an opponent faces the problem of when the resource for surprise should be exploited and when it should be conserved for a time at which the stakes are higher and the surprise would be more valuable. A classic example of a resource for surprise is the British control of all the German agents in Britain in 1942 (Masterman, 1972). The British recognized that the German intelligence system was vulnerable due to its heavy reliance on spies. For two years, the British waited to exploit their ability to mislead the Germans, until D Day, when the stakes were very large. Their patience was amply rewarded. False information given simultaneously to all the spies produced for the Germans a correlated shock. The message from each of "their" spies reinforced the credibility of the others. The Germans fell for the grand deception and kept a large number of troops at Pas de Calais—even several days after the real attack at Normandy.

There are several important implications of the fact that information systems may be attacked precisely when the stakes are very high (Axelrod, 1979). First, for the attacker, patience is a virtue since it may pay to exploit surprise by waiting for rare events with very large stakes. Second, for the defender, it would be a mistake to evaluate the risk of being surprised by what was seen when the stakes were low or moderate. Actual or potential opponents may be waiting for an opportunity of sufficiently large stakes to justify the exploitation of whatever resource for surprise they may have. Thus, judging the reliability of a spy, or the reliability of a crucial computer system, by its performance in a series of relatively low stakes circumstances could be quite misleading. Third, when the stakes get very large, the risk of being surprised is greatest.

Immunology provides some valuable insights in the resistance of a Complex Adaptive System against malevolent attacks. In the case of attacks by pathogens, the mammalian immune system is able to protect the host by distinguishing between foreign material and self. Distinctive protein patterns serve as tags that permit immune system cells to identify what is part of the self. Experience with particular pathogens often results in immunity against further attacks of the same or similar kind. Conversely, the human populations that have been the most vulnerable to disease are those that have been isolated on continents or islands and then have suddenly become exposed to pathogens that are new to them (McNeill, 1976).

It has become clear through the term "computer virus" that considerations of immunity also apply to information systems. A reasonable speculation is that information systems that have been exposed to numerous attacks from hackers have had many of their weaknesses exposed and corrected. Conversely, information systems that are isolated may actually be more vulnerable to attacks if they ever do become exposed. There are two policy implications for information security. First, the effort to protect critical information systems by isolating them may actually make them more vulnerable if their isolation cannot be guaranteed. Second, if security is to be achieved through redundancy, it can help to have the redundant systems be as different as possible (rather than exact copies of each other) so that some system might be able to resist an attack that is fatal to the others. This second principle has played a central role in the research of our colleague Stephanie Forrest, who has worked to devise immune systems for computers as an alternative to standard approaches that vaccinate against identified threats. She has shown promising results for systems that can uniquely tag their own processes so that they can distinguish self from other in order to identify attacking programs without the attackers having to be previously identified (Forrest et al., 1996; D'haeseleer et al., 1996).

Organizing Routines

The distribution of stress is not the only way that interactions within the system stimulate further activation of agents. Another important mechanism of this kind is the formation of work routines in organizations. Work routines also are recurring patterns of interaction among agents and artifacts. Because routines combine the distinctive skills of multiple human agents, the interactions in a single routine may be quite diverse. A chain of individuals assembling a car can do many more things than falling grains of sand can do. However, the basic mechanism of stimulating further activation works in a similar fashion. Routines arise because interactions among agents increase the likelihood of later repetitions of those same interactions. Usually this happens through learning by the participants. They may become aware of a valued result from an overall routine in which their actions played a part. (An emergency room worker hears that the patient who was referred to cardiology three days ago has been able to go home. A referral in future cases that are similar becomes more likely.) Or an agent is aware that some appropriate action has followed from a step previously taken. (The next worker on the line more easily processes the part a colleague carefully positioned. In the future, that same pre-positioning will be used.)

We normally do not give too much thought to how routines arise. They are important sources of organizational productivity, but part of their value rests in accomplishing work while taking relatively little attention. So routines are noticed mainly when they do not work, when they resist needed change, or when they "fire off" inappropriately (Nelson and Winter, 1982; Cohen and Bacdayan, 1994).

The way routines form is important for organizations. The easier it is to create good ones and modify bad ones, the more productive organizations can be. Various process improvement methodologies, such as Total Quality Management and Business Process Reengineering, have flourished in recent years in recognition of this fact.

The quality movement, in particular, has offered many procedures that make the linkage between events clearer to participants in order to make routines easier to learn and to improve.

One of the most famous devices of the early days of quality improvement was the system of cords that allowed Toyota workers to stop a whole production line when a defect was noticed. Tracing the defect to its upstream cause, rather than patching it locally, allowed all the participants to understand the interdependencies of the production routine. It is an expensive remedy, especially at first, when it is used often. But it is hard to imagine that Japanese auto manufacturing could have attained its reputation for quality without something like it. And the postwar world would be very different if that had not happened (Womack et al., 1990).

Looking at routines in this way, one sees that there are many devices for making the next step or the final result more visible to participants. These include feedback on total daily production, notices of receipt, in-boxes and out-boxes, periodic account summaries that report recent changes, even procedures to highlight the absence of feedback or complaint.

It is important to note that all of these examples of propagating stress and of self-sustaining activity are about formation: of avalanche potentials, of boundaries, of recurring action cycles. They are not about the observed structures but about how those structures arise. When we look for insights into harnessing complexity, we should ask how we can change the pattern of avalanche (or traffic jam) sizes, the shape and size of patches that form, or the number and complexity of routines that can be created. The theories often do not give us control over specific events. Rather they help us find interventions that may affect the averages of what happens, that may allow adaptation or learning, even without knowing in advance just what will change, or just what will be learned.

Restructuring of Physical and Conceptual Spaces

Our final mechanism of internal change in interaction patterns deals with space. In this mechanism the actions occurring within the system alter the very structure of the space in which actors are located. The agents are not directly intent on changing the collective interaction patterns, but barriers are being created (or reduced) from the inside, as a by-product of agent actions.

The classic example from biology is speciation. Over time a population can diverge, its members evolving into subgroups that eventually can no longer interbreed. The subpopulations have grown far apart, as though they were continents, now separated by a kind of ocean in the space of possible animals. The animals may have had no intention to form separate species, but their breeding decisions eventually had that result.

Biological examples of mergers are less common, but they do exist. One widely supported theory suggests that mitochondria, the tiny "fuel factories" of animal cells, are the result of a capturing process (Margulis, 1981). Bacterial structures were incorporated by host animal cells, and some of the genetic encoding for mitochondrial reproduction was moved to the animal cell genome. What had been a separate population was merged into another one.

In the social world, we frequently see merging and division of groups and even nations. At the level of national politics, such processes always have an explicit component. New nations declare their independence. Foreign governments recognize their existence. But frequently this is a late stage of what began as a more implicit and internally driven separation process. A group of people that have been considered part of some larger nation find themselves interacting more strongly with each other and less with members of the "other" group. They begin to talk of their separate identity. That talk, and the reaction of others to it, may then propel the dynamic into a new phase, one better understood with tools from the early part of our chapter, on external intervention and introduction of explicit barriers.

These remarks on the mixture of internal and external processes in nation building serve as a reminder that actual situations typically involve many mechanisms at once. Variety may be created by imitating foreign visitors even as it is destroyed by censoring media. Interaction may be decreased by a policy decision to reduce foreign language instruction, at the same time as interaction is increased by the need to trade with each other. An idea can be widely adopted in spite of being publicly condemned—indeed, *because* of being condemned, so that publishers will be eager to print a book "banned in Boston."

Having reviewed this collection of mechanisms for changing interaction patterns, we can now focus on the third major aspect of Complex Adaptive Systems, their mechanisms of selection. Interaction among agents shapes the creation and destruction of variety and produces the events that drive the attribution of credit. Now we can examine how selection itself works and see how it feeds back onto variety and interaction.

IV

Selection

In previous chapters, we have looked at the mechanisms that create and destroy types, and at the processes and structures that govern interaction among types. In this chapter, we turn to the fundamental question of which agents or strategies should be copied and which should be destroyed. In other words, how should selection be employed to promote adaptation?

Natural selection in evolutionary biology provides a familiar and well-studied example of how selection can work. Although selection in a Complex Adaptive System need not operate in the same way as natural selection, evolutionary biology is a good place to start our analysis. Evolution by natural selection requires three things. First, it requires a means to retain the essential character of the agent. In biological systems, genetic material preserves the key patterns. Evolution by natural selection also requires a source of variation. In the simplest biological systems, this can be achieved by mutation. In sexual reproduction, novelty is generated through recombination of characteristics from different parents, as well as by mutation. Finally, evolution requires amplification, changes in the frequencies of types. In biological systems this is the result of some individuals having many offspring while others have few or none.

If you want to design a system that is able to explore new possibilities while being able to exploit what has already been achieved, biological evolution provides an important benchmark. It demonstrates that adaptation can be achieved even without the agents (or anyone else) having any understanding of how the system works.

While natural selection provides an important paradigm for how an adaptive system can work, it also has some serious disadvantages compared with more directed methods of achieving adaptation. Whenever it is feasible to attribute success to something more specific than the entire agent, there is the possibility of selecting strategies rather than whole agents. If you find that quinine-related compounds reduce malaria, you can spread them through the world instead of waiting many generations for natural selection to breed malaria-resistant humans. This is especially valuable since the main antimalarial solution nature has so far evolved makes the carrier susceptible to sickle-cell disease, itself a debilitating condition. When attribution is sufficiently precise—and this can be far from perfectly accurate—it can pay handsomely to make numerous copies of a good strategy on a fast time scale that would be impossible if complete agents had to be reproduced.

These two approaches, selecting at the level of entire agents and selecting at the level of strategies, share the need to make copies that retain effective adaptations, to incorporate variation for further adaptation, and to amplify the success (and cull the failure) that does occur. But they differ in the level at which they operate—and selection at the two levels can work *very* differently. Selection of one advertising agency from a population of competing firms can have quite different dynamics from selecting among a population of advertising themes proposed by a single agency. Nonetheless, whether it is whole agents or strategies that are evaluated and undergo reproduction, a design for an adaptive system of selection must deal with four issues:

1. Defining criteria of success.
2. Determining whether selection is at the level of agents or strategies.

3. Attributing credit for success and failure.
4. Creating new agents or strategies.

This chapter will consider each of these elements of a selection process in turn. While these elements do not separate neatly in the everyday world, distinguishing them will help simplify our discussion without introducing too much distortion.

The chapter includes two extended examples that show how principles for harnessing complexity can be applied. The first deals with improving the criteria for success used by agents. It explores prize competitions as a mechanism that changes success criteria in a field by identifying and rewarding exemplary individuals or activities. The second example suggests approaches to attributing credit for success and failure when there is only a limited amount of relevant experience available. It explores military use of simulation as a method of generating surrogate experience that can be helpful in accelerating the attribution of credit to strategies that might be successful. The chapter concludes with a discussion of how leaders can use their visibility to help shape the selection of effective agents and strategies.

Defining Criteria of Success

The importance of knowing what to count as success is the point of an Army story about the new draftee who was an operations analyst in civilian life. After standing with fellow draftees in a long line to get their dinner plates washed and rinsed, the recruit went up to the old sergeant and explained it is inefficient to use two vats for washing dishes and two vats for rinsing them. It would be faster to use three vats for washing and only one for rinsing since washing takes more time than rinsing. The old sergeant looked with disdain at the new recruit, and said, "You've got it exactly backwards. I want them to stand just as *much* as possible. I can't keep them running around all day, but the longer I can keep them on their feet, the better."

Clearly, selection of agents or strategies implies some metric of success. Agents need not attend to the measure. Animals can have many offspring out of motives far more compelling than the eventual adaptation of their species. Fashions may be copied without much big-picture reflection by those adopting a new style. In such cases, success is actually defined by outside observers as "frequently copied." Rather than specifying a success measure and copying what scores well, this approach measures success by numbers of copies. Biologists take this line when they assess fitness as number of offspring. In biology, survival defines what is fittest.

However, in most of the situations we consider, performance measures are active in the minds of designers, policy makers, and other actors, whether they are acting inside the system or contemplating it from the outside. Recall the example of Linux software development, with its thousands of volunteers proposing solutions to specific problems in a massive operating system. Being able to evaluate the effectiveness of proposed solutions using clear measures such as speed and crash-avoidance was one of the requirements for such open software development to work. Typically, however, the assessment of alternatives in a Complex Adaptive System is not easy. In fact, there is usually more than one criterion that could be used to assess results.

For a business, profit seems a natural measure of success. For a checkers player, winning games is a natural performance measure. Yet even in these examples, with success criteria that seem indisputable, complexity might be harnessed more effectively if other measures of success are used. In the business example, market share provides an additional measure that can be a useful supplement to profits. One reason is that changes in profits may reflect factors beyond the control of the company, such as improvement in the national economy. You might not want to attribute credit for increased profits to your new marketing campaign if you knew that your entire industry had prospered during a buoyant economy. An increase in your market share could provide a better indication than profits of whether you were doing something right—and what it was. We will also see below for our checkers example that there are measures of success that may be more effective than waiting for the outcome of the game.

Our approach to harnessing complexity does not take any performance measure as "given." It does not anoint any one measure as a highest goal. Performance measures can be seen as instruments that shape what events are likely to occur. Even the preservation of life is not a goal that trumps all others, as human willingness to die for principles so dramatically reveals. Since goals are not seen as fixed, setting goals, the criteria that govern processes of selection, is one of the main interventions for those who would harness complexity. Our view leads to two important and uncommon observations about performance measures.

First, it is valuable to appreciate that *performance measures are defined within the system*. They are modified (or maintained) and applied (or disregarded) by the agents themselves. This observation is not a surprise to many experienced practitioners, who are well aware of the political work that lies behind measures later taken as givens. Unfortunately, many efforts to apply complexity concepts to social systems give little attention to how performance measures are defined within the system. To see what we mean, consider the case of profit as one such measure. What may count as a profit depends on many factors, including what the law allows individuals to own, what social norms and religions define as morally fair, whether actual practices conform to those norms, what the tax code recognizes as legitimate costs, and whether society charges for disposal of the by-products of activity, such as used motor oil or even carbon dioxide. We also regulate the scope of profit as a permissible goal. We largely removed profit from the decision making within American schools, hospitals, and prisons at the beginning of the twentieth century and are experimenting now with reintroducing it.

A further consequence of performance measures being defined

by the agents themselves is that there can be more than one measure active. In addition, the measures may be inconsistent and may change over time. Change that is seen as improvement by one type of agent may be seen as a loss by others. There are issues of variety in performance measures just as there are in other characteristics of agents and their strategies. When members of an organization assess a situation from different evaluative angles, they generate a greater variety of new possibilities that, if not excessive, can have great value for the organization (Cohen, 1984). But it is clear that beyond some level, variety in performance measures can also be a source of debilitating inconsistency and conflict.

Second, *how success is defined affects the chances for effective learning.* To return to our example of checkers, consider the difficulties for learning if victory is the sole criterion of success. The central problem is that victory or defeat comes only once per game. However, getting more than one measurement of performance per game could dramatically improve the rate of adaptation. The typical way to do this is to use criteria that can be measured in the course of the game. In checkers or chess, this is possible by evaluating the current board to see who is ahead in pieces and in various aspects of position. Such evaluations allow intelligent choices in the midst of the game based on what promises to lead to a better board position in a few moves. This doesn't require seeing all the way to victory or defeat at the end of the game. Since you cannot precisely measure the consequences of early moves for victory, you introduce other metrics that are more easily predicted. In a seeming paradox, you increase the chance of winning by concentrating on a set of criteria that does not include winning.

Even better, with finer-grained measures you can actually learn to improve the criteria by which you evaluate board positions. For example, you might learn from experience that having many pieces in the center often leads to surprisingly good results a few moves later. Indeed, the Samuel checker-playing program, one of the early triumphs of artificial intelligence research, learns on its own to play better checkers by using expected results in just this way (Samuel,

1959). When it arrives at a board position that is surprisingly good or bad, it uses this information to revise its own success criteria. The program determines what changes in its evaluators would have avoided the surprise and makes the corresponding changes. When it next encounters a similar board, the program will have a better set of criteria for attributing value to board positions. This approach to learning new success criteria is very powerful. Samuel's program, running on an early computer that could not keep up with today's digital wristwatches, could learn checkers well enough to defeat a state champion. Moreover, these are techniques of very broad applicability:

- When success is measurable only rarely, new measures with a faster tempo can speed learning, even if they do not perfectly reflect the longer-term goal.
- Whenever outcomes are better or worse than expected, the experience can help to revise evaluation criteria so that, in the future, the attribution of credit will produce better outcomes (Cohen and Axelrod, 1984).

Using fine-grained and short-term measures of success can help individual learning by providing focused and rapid feedback. Such narrow and prompt measures of success can also be used by an organization to evaluate who is successful and who is not. For this reason, managers are often judged by how well their unit does each quarter, or even each month, or by very specific indicators such as cost reductions. But there is a problem. If the challenges the manager is dealing with are long-term or widespread in the organization, then using fine-grained and prompt measures of success can easily miss much of the value to the organization of any improvement the manager discovers. As we saw in Chapter II, on variation, there can be a lot of bang for the exploratory buck when advances in one domain can be applied for a long time and/or in many places. A challenge for an organization is to develop measures of success that support appropriate levels of exploratory behavior while taking into account that learning is fostered by fine-grained and rapid feedback.

Another challenge in defining measures that will support learning is that a measure may be correlated with what ultimately matters without actually being causally related. A medical example is the reduction of fever as a measure of success in fighting a disease. A fever indicates the presence of a disease, and the fever disappears when the disease does. But with the development of aspirin, one can reduce the fever without curing the disease. Therefore, using body temperature as a measure of success can be misleading for some diseases. Because the elevated temperature might even be part of the body's method of fighting the disease (Nesse and Williams, 1998), parents may learn to treat the fever with aspirin in ways that can actually be harmful. The implication is that one needs to be careful about which indirect measures of success are used to guide action and learning.

Taken together, these observations about success measures imply not only risks but also rich possibilities for harnessing complexity through shaping the criteria by which the agents or their activities are evaluated. Performance measures are not immutably given, but are subject to change, both from the outside and from within the system where they operate. What measures are used profoundly affects which agents and strategies will be copied and recombined and, therefore, what adaptation will occur. This is the logic that gives long-term power to what may seem modest changes in measures, such as introducing on-time performance into airline regulation, body counts into battle assessments (Gartner, 1997), "pawn structure" into chess (Euwe, 1968), and portfolio risk into financial management (Sharpe et al., 1998).

Example: Prize Competitions

A particularly illuminating example of changing success criteria is the method of establishing a prize competition. Consider, for example, the ancient Athenian practice of conducting annual dramatic contests (Pickard-Cambridge, 1968). By explicitly declaring which drama was the best, the award accomplished three things. First, the author was honored for success, bringing fame and influence to individuals such as Aeschylus and Sophocles. Second, the award encouraged the production of new plays composed to meet the criteria implied by the previous awards. In our terms, the strategies of later playwriting were changed. Third, the award helped educate and shape the tastes of the audience, thereby providing future support for the criteria of excellence the award implied.

Today prize competitions are used to reward, encourage, and define excellence in a wide range of activities, from grammar school art contests to the Nobel Prizes in physics, peace, and literature. There are now prizes for beauty, for most valuable player, for best dressed, and for business quality. The effectiveness of prizes is enhanced as society develops more extensive channels to disseminate news of awards. So we should not be surprised that their use is increasing. Every increment in the reach of printing, television, or e-mail newsgroups increases the possibilities for affecting success criteria by announcing winners of awards.

Some prizes are for accomplishments that can be assessed more or less objectively, such as the winner of a solar-powered car race. For our purposes, the most interesting prizes are those that are based on subjective criteria. Indeed, for many prizes, the criteria are so indefinite that the burden of defining excellence within some realm falls heavily, if not entirely, on the subjective evaluations of a panel of judges.

From the point of view of harnessing complexity, a major advantage of prize competitions is that they can award credit to people or activities based on criteria that are different from current standards. The presumption is that a carefully selected panel of judges can make worthwhile evaluations of quality. The indirect effects are as powerful as the direct effects. Giving a prize not only rewards a winner who might not have excelled in other assessments but also provides a target for others to emulate. Emulation may take the form of superficial imitation, but it may also create innovative exemplars of just what was most valued by experts. In addition, by helping to shape the tastes of the general audience, a prize competition can also shape the criteria used by the broader public. For example, book awards not only provide guidance to writers and publishers about what is being valued but also provide guidance to readers and reviewers about what is worth reading. The promotion of a sophisticated reading public, in turn, helps provide a market for good writing.

A prize competition can also promote useful variety. Prizes sometimes serve to identify and promote things that are new and valuable. When a science or literary prize is awarded, it tends to legitimate and promote the entire field or genre of the winner. Of course, there is a tension here. Deciding who or what should receive an award involves the application of standards of excellence. The judges inevitably use standards that are shaped in part by the standards in the broader community of which they are a part. Indeed, judges are usually selected on the basis of their own standing, which in turn is often based on their adherence to current standards. And even if the judges may wish to be leaders in the identification of what is both new and worthy and are willing to take a risk on something that stretches current standards, they also need to be concerned about looking arbitrary or even foolish. The judges are also judged. Therefore, they face the familiar trade-off between exploitation and exploration in making their selections. The trade-off creates a tension between making a safe choice that reflects current standards and making a bold choice that can help transform those very standards.

To the extent that prize committees are willing to go beyond the orthodoxy of the moment, they represent a valuable potential for increasing useful variety. This potential is not always fully seized. Of the first 85 winners of the Nobel Prize for literature, all but one wrote in a European language (Espmark, 1986).

Prizes can even stifle variety. It is now very hard for a young pianist to establish a successful recording or concert career without having won one of the major competitions. The reason is that producers rely on the competitions to screen pianists. Young pianists therefore train to win these competitions, go to teachers who have won or whose students have won, choose repertoire suited to winning, and so on. Thus there is some truth to the criticism that competitions can reduce the variety of piano expression exactly because the competitions can become the dominant focus for young players. It can take a long time for the weak signals of public taste or music reviews to counter the now strengthened signals of prize jury standards.

While each prize sets up a competition among those aspiring to win it, there is also competition among the prizes themselves. The sponsors and judges of each prize seek attention and prestige for their award. Within each domain there is competition for how much credit will be garnered by the winners of a particular award. Is a Pulitzer Prize for fiction better than a National Book Award? Prize competitions themselves interact, as when getting one prize makes a winner more likely to get another prize. Moreover, a lesser-known prize can gain prestige if its winners often go on to receive some better-known prize. Thus there is an intricate set of interactions within and between four populations of agents: prize seekers, members of their audience, judges on awards panels, and the various prize competitions themselves. Together they function to alter the criteria that define success in their respective domains.

Determining the Level of Selection

Two basic processes amplify success: selection of agents and selection of strategies. The natural selection of biological agents works by making an entirely new agent without the need to determine the cause of the success of the parent or parents. The selection of strategies, on the other hand, creates new strategies for an existing agent. It often involves some explicit decision about what strategy or part of the agent was responsible for the success.

Selection of Agents

Biological systems are not the only ones that select entire agents. Elections are another such method. If a congressional representative is defeated in an election, another person gets the job. The voters are not able to pick and choose among the features they like in the incumbent and a challenger. They simply have to pick one candidate or the other. This provides an easy answer to the question of what should be given credit for success (or failure). The answer is the whole candidate. As much as a voter might want to give credit and blame separately for some good and bad policy positions or character traits, the vote requires selection at the level of the whole agent. One agent will occupy the office for the coming term; all others will be cast aside.

Elections offer a nice example of several coevolving complex systems. While voters are selecting at the level of agents, active politicians are selecting at the level of strategies. They observe carefully what positions were taken by recent victors around the country. Many will adopt those more successful strategies in future campaigns.

The economy also can select at the level of agents. Companies that go bankrupt and are liquidated are thereafter not present in the population. On the constructive side, imagine a decentralized firm that has a highly successful branch office. The firm might use its earnings to "clone" the successful branch office by setting up another branch that, insofar as possible, duplicates the entire operation of the successful one. If the branches operate fairly autonomously, this would amount to creating a new agent. The central office would have given credit to an entire branch (rather than to any of its particular strategies or characteristics) and tried to amplify success by producing a duplicate agent.

As we noted, biological evolution works by selecting agents. The success of an organism leads to reproduction. This does not entail any determination of which of the genes "deserve" credit for the reproductive success. Instead, all the genes in the reproducing organism get a roughly equal chance to be passed on to the offspring. This fact is the root of the phenomenon called hitchhiking, in which nonproductive, even mildly deleterious, genes are carried into subsequent generations by the success of the overall agent package to which they belong (Maynard Smith, 1978).

In all these examples of agent selection, there must be fairly substantial accumulations of resources to create a new agent, whether that agent is an infant organism, a political candidate, or a branch office. The need to accumulate sufficient resources to embody a new agent operates as an important limiting factor in agent-level selection. It contrasts with the situation we will see below for strategy-level selection, where what is copied can often be merely the abstract pattern of the strategy. The extreme example of this, a process that is profoundly reshaping our era, is the copying of computer algorithms. Here, the marginal costs of assembling a new copy may hover just above zero, allowing low-cost software to run on millions of computers.

When using selection of agents to harness complexity, a key question is how strong the **selection pressure** should be. If the best agent in a population gets many copies while the others get few or none, the selection pressure is very high. In effect, strong selection pressure greatly amplifies the success of the best agent in the population but gives very little amplification to the slightly less successful. In an era where franchising can provide strong selection pressure, the best ideas for a hardware store or a bookstore will be extensively copied, while independent competitors will languish. Conversely, weak selection pressure produces only a slight tendency for the better agents to have more copies and thus provides more uniform amplification to the relatively successful agents. The advantage of strong selection pressure is that it exploits success by quickly spreading copies of the best-performing agents. The disadvantage is that it can quickly destroy the variety in the population that is needed to explore for even better outcomes in the future. Thus the trade-off between strong and weak selection pressure raises the familiar issues of choosing the balance between exploiting the best current outcomes and using variety to explore for possible future improvements.

Managers and designers often have opportunities to change selection pressure. Among other things, they can increase rewards and visibility for top performers and set severe punishments for flaws. For example, "zero tolerance" deletion of agents or artifacts with small deficiencies has the effect of reducing variety. It thereby favors exploitation over exploration. In the short run, strong selection pressure converts existing variety to new exploitation, but in the long run exploration may suffer. Harnessing complexity requires taking advantage of variety rather than trying to ignore or eliminate it.

An instructive issue in biological reproduction is the **founder effect**. An example would be an island populated by long-beaked birds descended from a long-beaked pair that were among the first to reach the locale. In its early history, the population is small, and an outstandingly fit individual has offspring that form a large portion of the next generation. Over subsequent generations, many traits of that "founder" are carried widely through the population. Whether or not they make their own functional contribution, the traits that made the founder effective co-occurred with traits that do not have high value.

Both kinds are amplified. A nonbiological example can be seen in the Carnegie libraries that proliferated in the United States in the early twentieth century. Many different communities established libraries starting from the same plans. Overall, the favored plans were good ones and carried financial subsidies. The practice of using them was beneficial on the whole but did result in libraries with specific services that were arbitrary or even unwanted in some communities in which they were instituted (Van Slyck, 1995).

Selection of Strategies

An alternative to selecting entire agents as the basis for the amplification of success is to make copies or recombinations at the level of particular strategies. In this section, we examine three common points of difference in selection at the agent and strategy levels: cost, waiting time, and difficulty of inference. We show how the different strengths of the two levels of selection are sometimes complementary, and that there can be substantial advantages to a hybrid system in which selection goes on simultaneously at both levels.

If success can be assessed at the strategy level rather than the agent level, one difference that often occurs is a lowered *cost* of copying. To assemble or acquire a whole new agent (a new person, a new business, a new governmental unit) is typically more costly than to copy a strategy employed by a successful agent. It takes years to grow several Pacific yew trees for bark that provides cancer-fighting compounds for a single patient. A laboratory synthesis of the active chemical makes it available quickly to many thousands of patients. An owner of a baseball team can try to buy a star pitcher from another team. If the reason for success is that the pitcher is winning by throwing the forkball, it might be cheaper to teach the other pitchers that strategy during the off-season. Whether this will be promising or not depends on how easily the forkball can be copied. Is there a pitching coach for hire with success in teaching it? Or perhaps success depends on the uncanny similarity of the star's forkball and fastball motions. Then it may be necessary to pay the cost of acquiring the whole agent, with the entire complement of strategies, or of searching for another pitcher with a comparable package of skills.

A second difference that often occurs between the strategy and agent levels is *waiting time*. One could just think of this as a special case of higher costs, but it deserves a brief discussion of its own. Because assembling copies of agents is generally a larger task involving more resources, it typically takes more time than copying or recombining strategies. Even if the direct costs of agent copying were affordable, the indirect costs of delay might not be. For example, another company may have a proprietary process for manufacturing a part that goes into a product you are developing. It might be quite valuable to invent your own process for making the needed component, and plausible to create a division within your company to do it. It would lower your costs and let you tailor the part to your particular needs. But competitors are racing to the market for your own product. The delay while you create a capacity to make the part means falling behind in competition with them. So you license the existing process from its owner, copying that strategy not because of lower monetary costs but because of the value of elapsed time.

To highlight the speed at which strategies can change, consider a stock market. Agents watch changes in prices for information about what other agents believe. Thus the market has a recursive nature in which agents' expectations are formed on the basis of their anticipation of other agents' expectations. The result can be rapid bubbles and crashes. Simulations of markets as Complex Adaptive Systems demonstrate how high rates of exploration can generate these bandwagon effects and "market psychology" (Arthur et al., 1997).

Social mobilization is another arena in which agents' expectations are formed by watching each other's behavior. Again, the result can be very rapid change once a bandwagon begins. The fall of the Berlin Wall occurred with amazing speed once the initial demonstrations showed what was possible. As in a market, people formed their expectations on the basis of their anticipation of others' expectations. Once begun, a series of demonstrations set off a cascade of revised beliefs leading to irresistible levels of protest (Lohmann, 1994).

Markets and demonstrations illustrate how strategies can be selected very quickly. Typically, selection at the strategy level is faster and less costly than selection at the agent level. Nevertheless, these differences are tendencies rather than inevitable consequences. So, by way of counterexample, large corporations are often faced with new products from start-up competitors. They sometimes find it quicker to create new divisions or small spin-off firms to make a comparable product rather than modifying existing lines of activity to produce it. In effect, this is a case where agent creation may be faster than strategy copying. Although the differences we have mentioned are only tendencies, they are rooted in the added difficulty that is typical for creating full agents. Hence it is often important to compare possibilities for selection at the agent and strategy levels.

A third difference between selection at the two levels involves problems that commonly occur in inferring exactly what is to be copied. There are myriad ways that selection can go awry and incorrectly reward an agent or strategy that was not responsible for a success. Such failures plague selection at both levels. However, one important difference does occur. Agents are collections of strategies. Successful agents generally use strategies that are mutually compatible. The interaction among those strategies does not have to be understood if selection is at the level of the agent, copying all its strategies. Biological selection of whole agents capitalizes mightily on this fact, but so can identical replication of franchised business units. Selection at the strategy level generally demands higher quality of inference. How many of the agent's action patterns must be copied to replicate the success? Which ones? To obtain the same low defect rate as a rival firm, which of their quality control procedures should be emulated? Selection at the agent level, on average, is more **context preserving** than at the strategy level. In a Complex

Adaptive System, where many results derive from effects that multiply other effects, context preservation can work to retain and spread synergies that are not fully understood. We made a related point in the previous chapter when we observed that the longer time horizons of those in authority create a common context for coordinating the faster actions of those they supervise. There we were examining agents' activities. Here we are examining the selection that follows from their success or failure.

We have argued that there are tendencies for selection at the agent level to be more costly, slower in elapsed time, and more context preserving. The first two effects are often not wanted, while the last one frequently is. This can set up a tension in which a designer or policy maker who has some freedom to influence the level of selection may have to trade off the various factors.

To take an example: Suppose that you want to discourage a dangerous behavior such as violating crucial safety regulations. We have usually considered selection for positive traits, but here we can look at negative selection. At the level of *strategy,* selection may correspond to punishing the action pattern. Each detected instance of rule breaking could be heavily fined, for example. On the other hand, *agents* could be negatively selected in response to their violations. An offending employee could be suspended, transferred, or even fired. These forms of removal will make the agent less likely to be copied. Taking the agent out of circulation and making the effort of replacement typically costs more and takes longer than simply changing an agent's strategy. If the safety violation is integrated with other strategies—for example if the agent's entire work style used a set of methods now considered unsafe—simply punishing the violations may not discourage the behavior, so removal may be worthwhile. If the violations are more a matter of "fashion"—for example, not wearing a hard hat in order to look fearless—punishing the action itself may be the preferred approach.

Schemes to amplify success are nearly always imperfect. Selection at the level of agent and selection at the level of strategy are families of mechanisms that have somewhat complementary strengths. Agent selection often works on longer time scales—faster is not always better—and preserves variation and context. Strategy selection isolates key patterns that can be more easily and rapidly copied.

Thus it is not surprising that there are many hybrid systems, where selection is found to be operating at both levels in a single population of agents. Many species of birds and mammals seem to select at both the agent level, by conventional natural selection, and at the strategy level, by processes of cultural diffusion, which operate at a much faster time scale (Lumsden and Wilson, 1981; Cavalli-Sforza and Feldman, 1981). In the human case, cultural evolution is so rapid and effective that we tend to ignore the continuing operation of natural selection. At the other extreme, we often do not notice cultural aspects of an animal population. But close observation reveals striking cases, such as the English birds that discovered how to peck through foil milk bottle caps. Their discovery spread across the entire country within a few years (Hinde and Fisher, 1951).

Hybrid systems such as this have tremendous advantages. Herbert Simon has argued that they are so beneficial that we could expect biological evolution to create individuals with increased susceptibility to following strategies suggested by others (Simon, 1990). Even though this "docile" quality makes it possible to take advantage of individuals who possess it, that can be outweighed by the tremendous gains of adding cultural selection of strategies to natural selection of agents.

These observations on complementary strengths and hybrid selection systems have a cumulative implication. When there is room to alter selection processes, it can be wise to look for changes to the system that could diversify it, adding fast elements if its selection processes are slow. If the fast processes are not succeeding, it can pay to add slower elements that sustain a new context. As an example of adding fast elements, organizations that rely heavily on change through personnel turnover are often ripe for improved trading of employee "war stories." A series of failures in piecemeal importing of "best practices" might suggest bringing in a new supervisor experienced in how the various routines form an interlocking system. As with many other interventions we discuss, hybridizing selection processes is not guaranteed to be better, but it is often a beneficial focus of attention.

Attributing Credit for Success and Failure

At the beginning of this chapter we outlined four interdependent aspects of selection. So far we have examined two of them, how success criteria are defined and whether selection operates at the agent or strategy level. To complete the picture we must consider two more issues:

- how an agent uses a performance criterion to increase the frequency of successful strategies or decrease the frequency of unsuccessful ones, a step we call **attribution of credit,** and
- how agents or strategies that receive credit are copied, recombined, or destroyed.

Credit attribution, though difficult and necessarily imperfect, can nonetheless be designed to help harness complexity. In the preceding section, we pointed out that context preservation could be advantageous if the cause of apparent success is not fully understood. This indicates a general problem. Since Complex Adaptive Systems are inherently difficult to understand or predict, it follows that attribution of credit in selection will often be difficult and prone to mistakes. If it were feasible, the best response would be not to make mistakes in credit attribution. Because such mistakes can be very costly, vast bodies of academic knowledge and expensive social apparatus have been created to reduce them. Systems of logic, methods of statistics, and philosophies of science are all aimed at improving the extent to which our conclusions follow from our premises and evidence. There are public debates, professional review boards, and courts of law. All contribute to limiting the mistakes in attribution of credit that may drive selection processes.

Where these tools for improving inference are cost-effective, we certainly believe they should be used, and we applaud the work that maintains and extends them. However, despite all the effort put into these valuable resources, totally accurate attribution of credit is often infeasible. To see why, consider this list from a survey of research on the limits of rationality (Conlisk, 1996). It details the factors that make it easy to learn appropriate lessons from the experience accumulated in making a series of choices:

- clear rewards for the appropriate choices,
- repeated opportunities for observation or for practice,
- small deliberation costs at each choice so that frequent choices are easier,
- good feedback on the results of choices,
- unchanging circumstances that keep inferences valid, and
- a simple context that can be effectively analyzed.

The contrast of this list with the properties we have seen in Complex Adaptive Systems is stark. In complex systems, it is difficult to determine what should be rewarded or which choice is appropriate. Measurement of success is often infrequent, and shifting context makes few observations comparable. Deliberation costs for choices can be high, especially if they require the apparatus of formal logic or statistics, or social processes of choice such as scientific peer review—not to speak of court proceedings. Feedback is ambiguous. Circumstances, even goals, are changing. All of this follows from the fundamental premise: we are coping with systems that are complex and adaptive, not simple or static. In the short run we are not likely to have a direct approach that "gets it completely right." We will need as well the indirect methods of harnessing complexity.

The difficulty of attributing credit in real experience can be reinforced by considering a few examples. The war in Vietnam provides a striking case. Although war usually produces large rewards (and punishments) and, in the end, provides clear feedback on the result, none of the other circumstances for effective learning obtains. For the Americans, the Vietnam War was not a victory. But exactly what lessons should have been learned from it? There are many contending lessons and no obvious way to determine which candidates are most appropriate. Despite these impediments, lessons were learned by the American military. These included the need for decisive force in any future war, the need to avoid slow escalation, and the need to avoid civilian interference in the conduct of the war (Powell, 1995). These lessons—"strategies," in our terms—were applied to the planning and conduct of the Gulf War and seemed to be effective in that application.

On the other hand, for the Soviet Union the Vietnam War was a success. The lessons drawn from the war by the Soviet Union emphasized that their Vietnamese allies won because of their great will and courage, assisted by military aid from Communist nations (Zimmerman and Axelrod, 1981). These optimistic lessons would not have warned the Soviets about the dangers of their later intervention in Afghanistan.

Biological systems also face difficulties in attributing credit. Consider birds, which determine from experience the visual characteristics of dangerous predators. Their situation illustrates one of the many interesting complications of credit attribution in a Complex Adaptive System: exploitation by others. The method is **mimicry**— as when many species of moths evolve spots on their wings that resemble the eyes of larger predators. This works because the birds develop a "prediction" of danger from appearance and rely on it to avoid predators. One presumes that the birds' capability to associate certain appearances with danger, which is a mechanism for attributing credit, serves the birds well overall. But the moths can also exploit the birds' imperfect credit attribution to avoid being eaten. Once again coevolution increases complexity and inhibits prediction.

For a nonbiological example of the limits of credit attribution, consider the person who ends the year with the highest sales volume, receives a significant bonus, and is singled out to be emulated. Years later, more careful cost accounting may show that most of the sales actually lost money for the firm because of eventual refunds or support costs. The business literature is rife with stories of performance indicators that failed to capture important aspects of a complex setting. These misattributions may occur because of causal connections that no one understands, or because some employees, like the spotted moths, come to mimic features that other employees, like the birds, have come to associate with success or failure.

The difficulties of credit attribution are endemic in Complex Adaptive Systems. Our aim in this section is not to escape them, though we recommend that when it is feasible. Instead, our aim is to suggest how the side effects of inevitable mistakes of attribution can be turned to some advantage. Each of these three categories is constructed as a composite of actual cases in which complexity makes some mistakes of attribution inevitable. They illustrate three different problems of inference that are highly characteristic of credit attribution in complex systems:

- the mistake of crediting or blaming a part when a larger ensemble is responsible,
- the mistake of attributing credit or blame to a particular ensemble of factors when in fact a different ensemble is responsible, and
- the mistake of crediting a misconstrued strategy, where the action involved produced success, but the conditions in which the action should be taken have been misunderstood.

The first type of mistake, crediting a part when a larger ensemble is responsible, is very common in Complex Adaptive Systems since they so often involve a number of entangled causal factors. It is easy to notice that a single agent or strategy is associated with a series of successes (or failures). If you are not positioned to observe the operation of other necessary forces, you reach an incorrect conclusion that it alone causes the results. Consider a manager of a department that uses project teams assembled for specific tasks. If it is the practice in the unit to reward team members whose work contributed to notable success, a manager can almost be sure that there will be some occasions where an individual receives credit for what was produced by the interplay among contributions of several team members—what is sometimes labeled the group's "chemistry."

We have stressed insufficient exploration in examples throughout this book because we so often have seen variation being undervalued by managers of Complex Adaptive Systems. But for this case, let's stipulate that the manager believes the department has a problem of insufficient exploitation. Perhaps "back channel" communication suggests that a project group has done well by ensemble effect rather than the efforts of the most prominent individual. How can the manager get "mileage" out of discovering those attribution mistakes without knowing what caused them? One approach is to make a special effort to reassemble that identical team for a later problem, retaining (and exploiting) the uncredited ensemble that may be there. Doing this has a cost, of course. It reduces the ability to mix and match individuals to the characteristics of the next task. Harnessing complexity does not always come for free.

The second type of mistake, attribution of credit to the wrong set of factors, is often made in Complex Adaptive Systems for much the same reasons. Diagnosis of causes in complex, multicausal situations is error prone. We might take as an example the problem of examining customer complaints about product malfunction in order to discover product defects or possible design improvements. Many large consumer product companies have service desks that answer thousands of calls per week about products. They frequently have systems that generate "trouble tickets" associated with each call. It is natural to ask what can be learned from the records of all this work that would contribute to improvement of the products, but closing this loop of organizational learning has often proven quite difficult.

Working with a group of such reports, an analyst searches for patterns in the way the features and structure of the product interact with the circumstances of use reported by the customers. The hallmarks of complexity are present. The analyst may develop hypotheses such as: "All these customers reported that sound quality deteriorated when they were driving on country roads. Could it be that the audio unit is disturbed by shocks spaced at a particular frequency?"

Many hypotheses like this one are generated, but not all will be correct. In many organizations, such hypotheses are tested by checking if they are sufficient to reproduce the problem. In a complex world, many of those tests will fail. Someone from product development (not the same division as customer service) will subject the unit to low-frequency jolts and observe that it still performs well. An interesting strategy at such a moment of impasse is to bring into the process some of the frontline customer service agents who took the original calls. They may suggest something like, "These all came in last winter. Does it only happen if the unit is cold?"

Of course, this may not turn out to be the answer. But in an organization having trouble maintaining contact patterns between two divisions, the effort to correct a misattribution provides an occasion for interaction during which other useful information may flow. It functions as an episode of triggered recombination. Product people learn of other patterns noticed in customer service. The frontline agents learn about new product ideas in development and can then be alert to relevant remarks from customers.

Our third class of mistake is failure to appreciate the critical role of context. This kind of mistake is especially common when selection is at the level of strategies because strategies so often take the form of conditional action patterns: "If you encounter circumstances X, then do Y." The problem is that the actions are frequently much easier to observe than the conditions. For example, if your opponent in a chess game gives you the opportunity to take a piece, it may not be easy to determine from the context if this is a stupid blunder or a clever sacrifice.

To take another example, suppose you are building a collection of rare books. Bidding at book auctions may allow you to observe the buying actions of your colleagues. But if there is competition among the bidders, they may not be willing to fully, or accurately, disclose why they bought what they did, when they did. Competitive barriers to observation are often a serious impediment to strategy-level selection. Moreover, the ultimate effect of buying decisions may not be clear for some time. It can take a while to appreciate the effect on a collection of new additions, and the market for particular kinds of holdings may grow or decline.

In such an environment, learning will go slowly. Efforts to emulate apparently successful buying strategies will involve mistakes because so many factors determine the ultimate success of a purchase, and because inferences about the conditional part of the strategies are so constrained. As we noted earlier, it could be advantageous in such a situation, as in chess or checkers, to develop shorter-range measures of factors correlated with long-run success.

Again, we look for ways that the inevitable mistakes of credit attribution can provide opportunities to harness complexity. In this case, it may be possible to gradually identify signals observable in the short run that can foretell the long-term performance that is the ultimate goal. One good approach follows Arthur Samuel's insight into learning to play championship checkers. Surprises are actions that came out better, or worse, than expected. Either kind can fuel improvement. The essential thing is to see what factors were observable or predictable in the short run that were correlated with the surprise. This is a powerful idea that has been found to work not only in artificial intelligence systems but also in the neuropyschology of human learning (Cohen and Axelrod, 1984).

To return to our rare books example, we might ask what other copies of a target book were recently in the market? Are there details of its condition that might add to its value? Is the market for this type of book cyclical or sensitive to economic conditions? Are new categories of buyers entering the market who might prefer books of this type? There are hundreds of these factors, which is why it is very hard to learn to buy well for a collection. But the harnessing complexity approach does suggest an important shift in question, asking, "What observable criteria were often high or low when you did better or worse than expected?" The search is not for what predicted the outcome but for what predicted the surprise, the deviation of your expectations from what occurred. Those are the factors to which you should give increasing credit if you want to speed the process of learning which factors to credit.

Example: Military Simulation

The problems of inferring proper lessons (attributing credit) based on limited experience occur in almost every sphere of human activity. Because military organizations only rarely obtain feedback from actual combat, their circumstances make adaptation especially difficult. Since credit attribution has long been so problematic in warfare, military organizations have a rich history of refining various forms of simulation, including many forms of gaming and field exercises. The techniques used by the military to cope with the problems of credit attribution when feedback is scarce are therefore particularly illuminating.

For these organizations the problem of determining what works well is especially vexing. Large-scale fighting is infrequent—and much work goes into keeping this true. That means that opportunities to try new weapon systems or tactical concepts, or to test officer capabilities, come rarely. Learning only from real combat experience is an unacceptably slow strategy for improvement. This is a price society happily pays for peace, but it leaves military organizations facing a difficult learning problem.

Where a firm might have several different versions of a consumer product tested in the field within a few months, a military organization might not accumulate the equivalent amount of useful experience in several decades. For an extreme example, there has never been any full combat experience for our intercontinental ballistic missile hardware, operational concepts, and crews. (Robert Powell, who creates mathematical models of nuclear deterrence, says that this field is the only branch of science where success is achieved by never having any data.)

A large portion of what military organizations learn about new technology and operational concepts must come from various forms of simulated experience. These may be war games, field exercises, small-scale engagements, mental experiments, computer models, or even imaginative reconstructions of military history.

The Information Revolution is providing computer tools that dramatically expand simulation possibilities. The United States military now routinely employs simulated aircraft, tanks, ships, and soldiers in its investigations of combat possibilities. Mobilizations of large forces for field exercises incur substantial resource costs, and even without live ammunition, there are inevitable injuries and deaths from the risky movements of personnel and heavy equipment. Such exercises cannot be repeated many times in minor variations, although exactly this capability is extremely useful in exploring a Complex Adaptive System, where deliberate variation of multiple factors may reveal large consequences.

The value of these new possibilities is also becoming evident in the business world. Although useful experience is not as scarce as in the military case, there are many situations in which exploratory trials with the real system are not possible. Major reorganizations or changes of corporate strategy are like this. They often have huge costs, and if they don't work, they risk the bankruptcy of the entire firm. In response to this need, simulation tools for business decision making are beginning to appear. Firms are arising that specialize in building such simulation models. Some are spin-offs of computer gaming companies, while others have arisen from consulting practices (Farrell, 1998).

There are limitations, of course. One shortcoming is that simulations often place sharp and arbitrary limits on improvisation. While it is an extremely important source of military and business innovation, improvisation is generally not realistically supported in computer simulations, which often insist that the "players" obey rules and constraints that in real activities they might decide to violate. Although they may fall short of realism in significant ways, computer simulations provide the kind of rapidly assessed measure of success we have discussed previously. They generate only surrogate experience, but they can improve learning in an experience-poor domain if they are used wisely, with clear attention to their limitations.

Creating New Agents or Strategies

We have now examined three of the four aspects of selection processes set out at the beginning of this chapter: the definition of success criteria; the focus of selection on agents or strategies; and the attribution of credit that connects a measured success (or failure) to an agent or strategy. The fourth part of the process is the actual destruction of an existing agent or strategy, or the creation of a new one through copying or recombination.

We have already developed many of the key points on this topic in the course of our earlier chapter on variation, which focused on the closely related issues of creating and destroying variety. There we analyzed processes of copying and recombination, the occurrence of mistakes in those processes, and their contributions to the variation in populations. We also considered destruction of the instances of a type, up to and including extinction.

In this section, we reiterate the key role played in our framework by making, recombining, and destroying instances of agents or strategies. We add to our prior analysis by considering the consequences of the differences in detail among the many processes we have grouped together under this heading.

The Key Role of Copying

Notions of copying are central both to biology and to computer science, two disciplines that have contributed enormously to complex systems research. These two traditions do not have identical notions of copying, and the differences between them are reflected in our framework. The biological approach to making copies is much closer to our discussions of selecting at the level of agents. For most agent copying, material resources have to be assembled, and copies are made using the same materials that constitute the copied agents.

By contrast, copies as conceived in computer science concentrate on preservation of abstract form. This view corresponds more closely to our discussions of selection at the level of strategies. This alternative view of copies reaches an impressive level of abstraction in binary-encoded information that preserves its essential character across arbitrary embodiments. A digital recording of a Bach fugue is a series of "ones and zeros" that can be represented as spots of magnetism, pits in an optical disk, or a series of voltage pulses or light waves.

Both notions of copy have a place in our framework because the way copies spread through a Complex Adaptive System does not always conform to the patterns seen in natural selection. There can be adaptation, but through patterns that are not necessarily like those seen in biology. A computer virus can spread much faster than a successful physical virus. Within hours it can clog thousands of computers all over the world with copies of itself. Thus a computer virus is different with respect to both time and space. Being immaterial, it can spread incredibly rapidly, and it can spread through a space in which "nearby" machines are physically far away. A Complex Adaptive Systems framework needs to encompass much more than the biological cases, even if those have provided much of its inspiration.

Detailed Differences Among Generic Copying Processes

Just as the difference between copying strategies and agents matters, so too do the detailed differences among various copying processes. Imitating someone's method for making telephone charity requests is not an identical process to passing along a photocopy of a fund-raising letter. Both involve copying, but the former involves far more integration of a pattern into one's own behavior. Setting an example that triggers imitation is very different when the population comprises nation-states than it is when the population is made up of schoolyard playmates.

By calling many different processes "copying," it has not been our intention to deny the important differences of detail. Indeed, as we showed in Chapter II, the details have to be studied very closely. Errors and recombining processes depend on those details. And the character of the variation in the system is shaped by them in turn. Making fund-raising calls using your friend's method is much more of a recombination of strategies than is photocopying and forwarding of a funding request.

While the detailed character of copying processes is of great significance, it is also important to discuss copying processes in the aggregate. That makes clear the deep similarities among Complex Adaptive Systems. Our aim in discussing "copying" in general is to guide designers and policy makers to ask questions about how copies are made, and how destruction happens, for the agents and strategies in the systems they work with. We want to stimulate the recognition of many different kinds of processes as "copying," from duplicating computer files to replicating fast-food franchises. Once copying mechanisms are identified, the questioner will have knowledge of the important details that we cannot have. In this way, the framework aims to suggest fruitful questions.

Exercising Visible Leadership

We have given many examples of what managers, designers, and policy makers might do in a Complex Adaptive System. Virtually everything we've said about how to harness complexity can be regarded as advice about leadership. In this section, we focus on one particular aspect of leadership that deserves special attention: that what a leader does is especially likely to be copied by others.

Why would someone want to copy the visible behavior of a leader? In the ambiguous and hard-to-predict world of a Complex Adaptive System, agents often don't know what criteria of success they should use or how to evaluate the strategies they could select. This is especially important in an age of uncertainty and rapid change. When adaptive agents live in a rapidly changing environment, they tend to look to other agents to see which performance measures tend to work and which ones tend to fail. When agents are not able to predict the effects of various possible courses of action, they may resort to imitating the observable behavior of agents who seem to be successful, or who at least have more experience with the new environment (Cialdini, 1984). Imitating others who are successful or experienced is a form of implicit attribution of credit that certainly has its disadvantages. When features that are copied are only superficially relevant, the results can be wasteful or even comical. Nevertheless, following the practices of those with more experience or success is often a good strategy in an uncertain world.

There are three basic reasons a leader in a formal organization or other social system is especially likely to be copied. First, a leader can sometimes set standards that provide incentives for others to copy. Second, a leader's actions or performance measures are typically seen to be successful and hence worth emulating. Third, a leader may set an example that helps establish beneficial norms in a community.

Leadership in setting a standard can cause others to go along for their own reasons. Consider the case of Norway as a country that writes much of the world's maritime insurance. When the standards body in Norway set certain regulations for insuring oil platforms, the makers of oil platforms had an incentive to build in ways that met those standards. Thereafter other marine insurers tended to gravitate toward similar regulations (Stinchcombe and Heimer, 1985). Norway's regulations helped shape the industry in ways that led other maritime insurers to copy their visible behavior.

The emulation of a leader need not be based on a full understanding of how the emulation will help. Other agents may wish to emulate the actions or performance measures of a visibly successful leader in the hopes that what worked for the leader will work for them. A business leader who wishes to promote environmentally friendly production can, of course, make decisions that give high weight to environmental concerns. But if the firm is highly visible and is able to show that it becomes more successful because of its environmental practices and reputation, then a much more powerful dynamic comes into play. Imitation of the firm's performance measures by other firms creates a cascade that can transform an industry. Many forms of inspirational leadership work in this same fashion. For example, Gandhi's criterion of nonviolence was advanced throughout the world by the success its practitioners achieved in winning India's independence from Britain. Gandhi's leadership was successful in large part because he visibly embodied the very values he was advocating. This led others to emulate not only his tactics but also his values (Gardner, 1995).

Visible leadership can also be exercised by setting an example that helps establish beneficial norms in a community. In Complex Adaptive Systems, norms are often important regulatory mechanisms. Central monitoring and control can be difficult when many agent interactions are widely distributed across physical or social spaces. Criteria that the agents themselves apply are a very attractive alternative. Especially when they become internalized, norms regulate not through fear of consequences but through the belief that some actions are right and others wrong. This is extremely important when monitoring by central authorities is costly or intrusive. Moreover, once established, a norm can be reinforced and spread by dispersed agents who accept the norm and are willing to punish others who deviate from it (Axelrod, 1986). The Internet is a vast example of opportunities for one agent to exploit another from afar. The eventual character of its culture will be established in large measure by decisions made in the next few years, as significant and highly visible leaders promote the norms they will exemplify and expect others to enforce. The major providers of e-mail and chat facilities provoke widespread debates when they announce or modify positions on how they will handle unwanted advertisements or offensive language. The dialogs that occur build communities of users who may well enforce standards among each other more effectively than central authorities could hope to do.

We have examined five major aspects of selection: criteria of success, focus on selection at the level of agents or strategies, attribution of credit, mechanisms for creating new agents and strategies, and the exercising of visible leadership. In doing so, we have seen that each aspect has dense connections to issues of variation and interaction. Our central question in considering selection has been, "Which agents or strategies should be copied or destroyed?" But the answer to that question is clearly intertwined with the two major questions of our earlier chapters, "What is the right balance of between variety and uniformity?" and "What should interact with what, and when?" In our concluding chapter we bring together these three elements of our framework for harnessing complexity.

V

Conclusion

We began by asking what actions you should take in a world with many diverse mutually adapting players, where the emerging future is extremely hard to predict. We answered by providing a framework for harnessing complexity: using the dynamism of a Complex Adaptive System for productive ends. In this final chapter, we summarize the ideas we have presented and comment on how they can be applied in the design of improved organizations and strategies.

The Central Elements of the Framework

There are a dozen central concepts that do a great deal of the work in our Complex Adaptive Systems approach. To apply our framework to a new situation, it is essential to ask how each of them can be interpreted in that context. We list these central principal concepts here, along with brief definitions. In the following sections, we show how they all fit together, and then we demonstrate how the concepts can be converted into active questions about actual settings.

- **Strategy,** a conditional action pattern that indicates what to do in which circumstances.
- **Artifact,** a material resource that has definite location and can respond to the actions of agents.
- **Agent,** a collection of properties (especially location), strategies, and capabilities for interacting with artifacts and other agents.
- **Population,** a collection of agents, or, in some situations, collections of strategies.
- **System,** a larger collection, including one or more populations of agents and possibly also artifacts.
- **Type,** all the agents (or strategies) in a population that have some characteristic in common.
- **Variety,** the diversity of types within a population or system.
- **Interaction pattern,** the recurring regularities of contact among types within a system.
- **Space (physical),** the location in geographical space and time of agents and artifacts.
- **Space (conceptual),** the "location" in a set of categories structured so that "nearby" agents will tend to interact.
- **Selection,** processes that lead to an increase or decrease in the frequency of various types of agents or strategies.
- **Success criterion** or **performance measure,** a "score" used by an agent or designer in attributing credit in the selection of relatively successful (or unsuccessful) strategies or agents.

How the Elements Form
a Coherent Framework

We can give a highly compressed restatement of how this book portrays a Complex Adaptive System. This puts things very abstractly, of course. It is unlikely to mean much to an unfortunate browser who has opened to the back of the book to see what it is about. But for readers who arrived here via the preceding chapters, the restatement may serve as a summary of how the parts of the framework form a working whole.

Agents, of a **variety** of **types,** use their **strategies,** in patterned **interaction,** with each other and with **artifacts. Performance measures** on the resulting events drive the **selection** of agents and/or strategies through processes of error-prone **copying** and **recombination,** thus changing the frequencies of the types within the **system.**

What a User of the Framework Asks

To apply the framework, the user needs to determine the meaning of the central concepts in the setting at hand. Here we suggest a series of questions that can help guide the user in harnessing the complexity of a particular system:

- What are the strategies, agents, and artifacts in the system? What are the ideas, rules of thumb, routines, and norms that agents rely on as they act? What are the tools or resources that they rely on?
- What are the populations of agents in the system? In particular, who can copy strategies from whom?
- What can I observe about how the agents themselves classify other agents and artifacts into types? Do the agents have special labels for categories of other agents, or for kinds of tools, or resources?
- How might How might *I* classify them? What categories of agents and strategies will be most useful to me as a designer or policy maker for harnessing complexity?
- What processes of copying and recombining create and destroy the variety of types? What additional processes might serve copying and recombining functions? Does new information technology offer new possibilities?

- What interventions would usefully create or destroy variety? How do errors occur in current processes? Does the variety that results ever offer potential value? Could other variety sources have greater promise?
- What is the right balance between variety and uniformity of types within the system? Is exploration especially valuable because improvements can be widely applied and/or used for a long time? Conversely, is there a risk of disaster from trying a bad strategy?
- What are the patterns of interaction among types? Are some agents following others? Are there agents, or signals, that should be followed?
- What interventions would change the patterns of interaction (in ways that are likely to be useful to the system as a whole; to you, as the designer; or to you, as one of the agents)? Are there physical or conceptual neighborhoods of interaction that need help in forming, or that deserve to be disrupted?
- What criteria of success does the system use to select the types that become more (or less) common over time? Are there multiple criteria within the population? Is selection done by many agents, or only by a few? Do performance measures make systematic mistakes in attributing credit?
- Is selection acting upon agents or upon strategies? Or is the system a hybrid, with selection at both levels?
- How should the selection of agents or strategies be used to promote adaptation?

What a User of the Framework Can Do

To show how the Complex Adaptive Systems approach can be used to generate actions, we can revisit our eight extended examples. Each example suggests a kind of action that takes advantage of complexity.

- *Arrange organizational routines to generate a good balance between exploration and exploitation.* In our example of military personnel systems, we saw how the movement of young officers through successive posts of great variety allowed them to construct novel combinations of ideas, some of which developed into the major innovations of the next generation.
- *Link processes that generate extreme variation to processes that select with few mistakes in the attribution of credit.* In the case of the Linux operating system, it was possible to use Internet communications to increase massively the variety of proposed improvements to the software. This did not destroy the integrity of the operating system, because appropriate and mutually agreed upon performance criteria could be used by the thousands of collaborators—for example, execution speed and crash-resistance of the modified software.

INTERACTION

- *Build networks of reciprocal interaction that foster trust and cooperation.* One way to do this is to promote the informal associations that provide the basis of social capital. We saw that northern Italy's advantage over the South could be traced back a thousand years to guilds, religious fraternities, and "tower societies" for self-defense in the medieval communes. Today northern Italy benefits from rich networks of organized reciprocity and civic solidarity fostered by cooperatives, mutual aid societies, and neighborhood associations. Informal associations, such as choral societies in Italy and bowling leagues in America, support long-term goals such as economic productivity and democratic governance. But these delayed benefits can be hard to ascribe to the shorter-run actions that produce them, so care must be taken to aid the development of rich networks of engagement that build social capital.

- *Assess strategies in light of how their consequences can spread.* Our example of AIDS research showed that it is important to take into account that interactions do not happen at random. The AIDS example also showed how the proper assessment of a vaccine requires a population perspective, since a vaccine's value depends not only on its ability to prevent the disease in the recipient but also on its ability to inhibit transmission.

- *Promote effective neighborhoods.* Our discussion of theoretical research on "tags" in the Prisoner's Dilemma game showed that there can be tremendous gains from helping would-be cooperators interact more frequently. In the simulation studies, arbitrary numerical labels on the agents fulfilled this function, but in many other settings, physical locations or social signaling devices such as clothing can perform the same role.

- *Do not sow large failures when reaping small efficiencies.* Our analysis of four failure modes of large systems showed that there can be significant risks from efforts to increase local efficiencies by linking processes that were not previously connected. Excess demand can be shifted among linked power or computing systems, but if the wider system fails, it does so on a much larger scale. The risks may be worth the gains, and good designs can minimize the risks. But the risks should not be overlooked.

SELECTION

- *Use social activity to support the growth and spread of valued criteria.* Our example of prize competitions revealed that the processes of refining prize criteria, selecting judges, recruiting nominees, and publicizing winners can all serve to disseminate the underlying goals that motivated the creation of the prize.

The result of such activity is to increase the use of criteria embodied in the prize, which can sometimes be far more effective
than direct advocacy of the criteria.

- *Look for shorter-term, finer-grained measures of success that
  can usefully stand in for longer-run, broader goals.* By examining the use of simulation in military and business affairs, we
  found that there can be severe shortages of experience to drive
  the adaptation of Complex Adaptive Systems. Although it pays
  to be alert to the risks of misattribution, it can sometimes be
  valuable to find ways to get more experience quickly, even if it
  is of lower validity. Simulations can do this. So can short-run
  proxy measures, like material advantage or center-control in
  checkers or chess.

This review of our extended examples does not cover all the implications of our approach, but it does illustrate the framework's
wide range of application, and it reemphasizes many of our major
points. Taken together, the examples convey the overall strength of
the framework as a guide to action in complex settings in which
long-term consequences may be highly significant but can be very
difficult to predict.

What May Come of This Approach

Like all authors, we harbor many ambitions for our book, but we will
mention only two. While it is not our central goal, we hope there will
be useful feedback to the scientific study of Complex Adaptive Systems. Researchers can see original contributions in our treatment of
the critical role of nonrandom interactions in adaptation, our systematic contrast of biological and informational copying, and our analysis of credit allocation and measures of performance. We know that
developing the framework has led us to new questions in our own research. For example, how do interaction patterns affect cooperation

in a community of Prisoner's Dilemma players (Cohen, Riolo, and Axelrod, 1999), or how can e-mail establish new communication patterns in offices (Cohen and Axelrod, 1999)?

We also believe that a Complex Adaptive Systems approach has rich possibilities for bridging the gap that often separates "humanistic" and "hard science" approaches to social analysis. It allows scholars to take into account more of the dynamic and multilayered reality of our social world, in which it is so clear that "history matters." The approach reduces the extreme simplification required to get results in "hard" social science, simplification that scholars in other traditions may fairly find unacceptable. At the same time, the framework puts some real constraints on reasoning about social systems, increasing the internal consistency of arguments, making some formalization possible, and improving chances for precise observation. This reduces complaints about "fuzziness" that are often lodged by advocates of social *science.*

Our framework provides natural ways to analyze institutions and how they shape—and are shaped by—the actions of individuals. It can accommodate agents who are situated in a rich fabric of social interactions and whose preferences may change with their experience. It can include the formation of new kinds of actors, or the disappearance of others, rather than ignoring history by assuming stasis. All of these are important to consider in social analysis and difficult to incorporate into most current approaches.

Our *primary* hope in writing the book, however, is for its impact not on the world of research but rather on the world of practice. We hope to contribute a coherent approach to designing interventions in a complex world, as managers, software designers, and city planners must do. Ours is not the first step in this direction. We have pointe out some of our inspiring predecessors—including great figures w have provided key ideas, such as Adam Smith, Charles Darwin, Herbert Simon. They are among many, many others, past and sent. Nor is ours the last step. This is a period of extraor change in technology, social structure, and—correspond ideas. As the scientific frontier of complexity research new organizational implications will be revealed that build upon the rich yield that is already available.

We all must intervene in Complex Adaptive Systems daily. We all face situations where the classical approach of formulating alternative actions and their likely consequences assumes more understanding and predictive power than we actually have. Our framework shows how the accumulating scientific insights into variation, interaction, and selection fit together and can be used to harness complexity.

References

Anderson, Benedict. *Imagined Communities: Reflections on the Origin and Spread of Nationalism.* London: Verso, 1983.

Arthur, W. Brian, John H. Holland, Blake LeBaron, Richard Palmer, and Paul Tayler. "Asset Pricing Under Endogenous Expectations in an Artificial Stock Market." In *The Economy as an Evolving Complex System II,* edited by W. B. Arthur, S. Durlauf, and D. Lane. Reading, Mass.: Addison-Wesley, 1997.

Axelrod, Robert. "The Rational Timing of Surprise." *World Politics* 31 (1979): 228–246.

Axelrod, Robert, *The Evolution of Cooperation.* New York: Basic Books, 1984.

Axelrod, Robert. "An Evolutionary Approach to Norms." *American Political Science Review* 80 (1986): 1,095–1,111. Reprinted in Robert Axelrod, *The Complexity of Cooperation* (Princeton, N.J.: Princeton University Press, 1997).

Axelrod, Robert. "The Evolution of Strategies in the Iterated Prisoner's Dilemma." In *Genetic Algorithms and Simulated Annealing,* edited by Lawrence Davis. London: Pitman, and Los Altos, Calif.: Morgan Kaufman, 1987: 32–41. Reprinted in Robert Axelrod, *The Complexity of Cooperation* (Princeton, N.J.: Princeton University Press, 1997).

Axelrod, Robert. "The Dissemination of Culture: A Model With Local Convergence and Global Polarization." *Journal of Conflict Resolution* 41 (April 1997): 203–226. Reprinted in Robert Axelrod, *The Complexity of Cooperation* (Princeton, N.J.: Princeton University Press, 1997).

Axelrod, Robert, Will Mitchell, Robert E. Thomas, D. Scott Bennett, and Erhard Bruderer. "Coalition Formation in Standard-Setting Alliances," *Management Science* 41 (1995): 1,493–1,508. Reprinted in Robert Axelrod, *The Complexity of Cooperation* (Princeton, N.J.: Princeton University Press, 1997).

Axelrod, Robert and William Zimmerman. "The Soviet Press on Soviet Foreign Policy: A Usually Reliable Source." *British Journal of Political Science* 11 (1981): 183–200.

Bak, Per. *How Nature Works: The Science of Self-Organized Criticality.* New York: Springer-Verlag, 1996.

Baker, Wayne E. *Networking Smart: How to Build Relationships for Personal and Organizational Success.* New York: McGraw Hill, 1994.

Belew, R. K. and M. Mitchell. *Adaptive Individuals in Evolving Populations: Models and Algorithms.* Reading, Mass.: Addison-Wesley, 1996.

Bonabeau, Eric. "From Classical Models of Morphogenesis to Agent-Based Models of Pattern Formation." Santa Fe Institute Working Paper 97-07-063, 1997.

Bornstein, David. *The Price of a Dream: The Story of the Grameen Bank and the Idea That Is Helping the Poor to Change Their Lives.* New York: Simon and Schuster, 1996.

Brand, Stewart. *How Buildings Learn: What Happens After They're Built.* New York: Viking, 1994.

Brown, John Seely and Paul Duguid. "The University in a Digital Age." *Change: The Magazine of Higher Learning* 28 (July/August 1996): 10–19.

Burr, Chandler. "Of AIDS and Altruism: In Theory, a New Kind of Vaccine Could Halt the Epidemic." *U.S. News and World Report* (April 6, 1998): 60–61.

Burton, Christopher. "The Radio Revolution." Center for Information Strategy and Policy, Science Applications International Corporation, McLean, Va. 1997.

Bush, Vannevar. "As We May Think." *Atlantic Monthly* 176 (July 1945): 101–8.

Buss, Leo W. *The Evolution of Individuality.* Princeton, N.J.: Princeton University Press, 1987.

Cavalli-Sforza, L. L. and Marcus W. Feldman. *Cultural Transmission and Evolution: A Quantitative Approach.* Princeton, N.J.: Princeton University Press, 1981.

Cialdini, Robert B. *Influence: The New Psychology of Modern Persuasion.* New York: Quill, 1984.

Cohen, Michael D. "The Power of Parallel Thinking." *Journal of Economic Behavior and Organization* 2 (1982): 285–306.

Cohen, Michael D. "Conflict and Complexity: Goal Diversity and Organizational Search Effectiveness." *American Political Science Review* 78 (June 1984): 435–451.

Cohen, Michael D. "Artificial Intelligence and the Dynamic Performance of Organizational Designs." In *Ambiguity and Command: Organizational Perspective on Military Decision Making,* edited by J. G. March and R. Weissinger-Baylon. Marshfield, Mass. Pitman Publishing, Inc., 1986.

Cohen, Michael D. and Robert Axelrod. "Coping With Complexity: The Adaptive Value of Changing Utility." *American Economic Review* 74 (1984): 30–42.

Cohen, Michael D. and Robert Axelrod. "Complexity and Adaptation in Community Information Systems: Implications for Design." In *Proceedings of the Kyoto Meeting on Social Interaction and Communityware,* edited by Toru Ishida. Heidelberg: Springer-Verlag, 1999.

Cohen, Michael D., Rick L. Riolo, and Robert Axelrod. "The Emergence of Social Organization in the Prisoner's Dilemma: How Context-Preservation and Other Factors Promote Cooperation." Santa Fe Institute Working Paper 99-01-002, 1999.

Cohen, Michael D. and Paul Bacdayan. "Organizational Routines Are

Stored as Procedural Memory: Evidence From a Laboratory Study." *Organization Science* 5 (December 1994): 554–568.

Coleman, James S. "Social Capital in the Creation of Human Capital." *American Journal of Sociology* 94 (1988): S95–S120.

Conlisk, John. "Why Bounded Rationality." *Journal of Economic Literature* 34 (1996): 669–700.

Cowan, George A., David Pines, and David Meltzer. *Complexity: Metaphors, Models, and Reality* (Reading, Mass.: Addison Wesley Longman, 1994.

David, Paul. "Clio and the Economics of QWERTY." *American Economic Review* 75 (1985): 332–335.

Dawkins, Richard. *The Selfish Gene*. Oxford and New York: Oxford University Press, new edition 1989.

D'haeseleer, P., S. Forrest, and P. Helman. "An Immunological Approach to Change Detection: Algorithms, Analysis, and Implications." In *Proceedings of the 1996 IEEE Symposium on Computer Security and Privacy* (1996).

Doyle, Roger. "Languages, Disappearing and Dead." *Scientific American* (March 1998): 26.

Eisenstein, Elizabeth. *The Printing Revolution in Early Modern Europe*. Cambridge: Cambridge University Press, 1983.

Espmark, Ejell. *The Nobel Prize in Literature: A Study of the Criteria Behind the Choices*. Boston: G. K. Hall, 1986.

Euwe, Max. *The Development of Chess Style*. New York: D. McKay, 1968.

Farrell, Winslow. *How Hits Happen*. New York: Harper Business, 1998.

Forrest, S., S. A. Hofmeyr, A. Somayaji, and T. A. Longstaff. "A Sense of Self for Unix Processes." In *Proceedings of the 1996 IEEE Symposium on Computer Security and Privacy* (1996).

Frank, Robert H, and Philip J. Cook. *The Winner-Take-All Society*. New York: The Free Press, 1995.

Gardner, Howard. *Leading Minds: An Anatomy of Leadership*. New York: Basic Books, 1995.

Gardner, Martin. "Mathematical Games: The Fantastic Combinations of John Conway's New Solitaire Game 'Life.'" *Scientific American* (October 1970): 120–123.

Gartner, Scott S. *Strategic Assessment in War.* New Haven, Conn.: Yale University Press, 1997.

Gell-Mann, Murray. "What Is Complexity?" *Complexity* 1 (1995): 16–19.

Gladwell, Malcolm. "Six Degrees of Lois Weisberg." *The New Yorker* (January 11, 1999): 52.

Gladwell, Malcolm. "The Science of the Sleeper: Collaborative Filtering Technology's Effect on Book Sales." *The New Yorker* 75 (October 4, 1999): 48ff.

Gleick, James. *Chaos: Making a New Science.* New York: Viking, 1987.

Granovetter, Mark S. "The Strength of Weak Ties." *American Journal of Sociology* 78 (1973): 1,360–1,380.

Hill, W., L. Stead, M. Rosenstein, and G. Furnas. "Recommending and Evaluating Choices in a Virtual Community of Use." In *Proceedings of the Conference on Human Factors in Computing Systems,* CHI '95 (Denver, Colo.: Association for Computing Machinery, 1995): 194–201.

Hinde, R. A. and James Fisher. "Further Observations on the Opening of Milk Bottles by Birds." *British Birds* 44 (December 1951): 393–396.

Hofstadter, Richard. *Social Darwinism in American Thought.* Boston: Beacon Press, rev. ed. 1955.

Holland, John. *Adaptation in Natural and Artificial Systems: An Introductory Analysis With Applications to Biology, Control, and Artificial Intelligence.* Cambridge, Mass.: MIT Press, 1st. ed. 1975; 2d. ed. 1992.

Holland, John. *Hidden Order: How Adaptation Builds Complexity.* Reading, Mass.: Addison-Wesley, 1995.

Holland, John. *Emergence: From Chaos to Order.* Reading, Mass.: Addison-Wesley, 1998.

House, James S., Karl R. Landis, and Debra Umberson. "Social Relationships and Health." *Science* 241 (July 29, 1988): 540–545.

Jacquez, J. A., J. S. Koopman, C. P. Simon, and I. M. Longini. "Role of the Primary Infection in Epidemics of HIV-Infection in Gay Cohorts." *Journal of Acquired Immune Deficiency Syndromes and Human Retrovirology* 7 (1994): 1,169–1,184.

Johnson, George. "Researchers on Complexity Ponder What It's All About." *New York Times* (May 6, 1997): B4.

Kahneman, Daniel and Amos Tversky. "On the Psychology of Prediction." *Psychological Review* 80 (1973): 251–273.

Kahneman, Daniel and Amos Tversky. "Prospect Theory: An Analysis of Decision Making Under Risk." *Econometrica* 47 (1982): 263–291.

Kauffman, Stuart. *The Origins of Order: Self-Organization and Selection in Evolution.* New York: Oxford University Press, 1993.

Kauffman, Stuart. *At Home in the Universe: The Search for Laws of Self-Organization and Complexity.* New York: Oxford University Press, 1995.

Kelly, Kevin. *Out of Control: The New Biology of Machines, Social Systems, and the Economic World.* Reading, Mass.: Addison-Wesley, 1994.

Kennedy, Bruce F. et al., "Social Capital, Income Inequality, and Firearm Violent Crime." *Social Science and Medicine* 47 (1998): 7–17.

Koopman, J., J. Jacquez, C. Simon, B. Foxman, S. Pollock, D. Barth-Jones, A. Adams, G. Welch, and K. Lange. "The Role of Primary HIV Infection in the Spread of HIV Through Populations." *Journal of A.I.D.S.* 14 (1997): 249–258.

Koza, John R. *Genetic Programming II: Automatic Discovery of Reusable Programs.* Cambridge, Mass.: MIT Press, 1994.

Koza, John R. *Genetic Programming: On the Programming of Computers by Means of Natural Selection.* Cambridge, Mass.: MIT Press, 1992.

Langton, Chris, editor. *Artificial Life.* Reading, Mass.: Addison-Wesley, 1988.

Lave, Jean and Etienne Wenger. *Situated Learning: Legitimate Peripheral Participation.* New York: Cambridge University Press, 1991.

Lloyd, Seth. "Physical Measures of Complexity." In *1989 Lectures in Complex Systems,* edited by E. Jen. Redwood City, Calif.: Addison-Wesley, 1990.

Lohmann, Susanne. "The Dynamics of Informational Cascades: The Monday Demonstrations in Leipzig, East Germany, 1989–91." *World Politics* 47 (1994): 42–101.

Luca Cavalli-Sforza, Luigi and Marcus W. Feldman. *Cultural Transmission and Evolution: A Quantitative Approach.* Princeton, N.J.: Princeton University Press, 1981.

Lumsden, Charles J. and Edward O. Wilson. *Genes, Mind, and Culture:*

The Coevolutionary Process. Cambridge, Mass.: Harvard University Press, 1981.

MacArthur, R. H. and E. O. Wilson. *The Theory of Island Biogeography.* Princeton, N.J.: Princeton University Press, 1967.

March, James G. "Bounded Rationality, Ambiguity, and the Engineering of Choice." *Bell Journal of Economics* 9 (1976): 587–608.

March, James G. "Exploration and Exploitation in Organizational Learning." *Organization Science* 2, no. 1 (1991): 71–87.

March, James G., Lee S. Sproul, and Michal Tamuz. "Learning From Samples of One or Fewer." *Organization Science* 2, no. 1 (1991). Reprinted in *Organizational Learning,* edited by Michael D. Cohen and Lee S. Sproul. Thousand Oaks, Calif.: Sage Publications, 1996.

Margulis, Lynn. *Symbiosis in Cell Evolution.* San Francisco: W. H. Freeman, 1981.

Masterman, J. C. *The Double-Cross System in the War of* 1939–1945. New Haven, Conn.: Yale University Press, 1972.

Maynard Smith, John. *The Evolution of Sex.* Cambridge: Cambridge University Press, 1978.

Maynard Smith, John and Eors Szathmary. *The Origins of Life: From the Birth of Life to the Origin of Language.* New York and Oxford: Oxford University Press, 1999.

McNeill, William H. *Plagues and Peoples.* Garden City, N.Y.: Anchor Press, 1976.

McNeill, William H. *The Pursuit of Power.* Chicago: University of Chicago Press, 1982.

McNeill, William H. "Information Technology and the Sweep of History." Lecture at Highlands Forum IV on Conflict in the Information Age. Santa Fe, N. M., August 12, 1996.

Milgram, Stanley. "The Small World Problem." *Psychology Today* 2 (1967): 60–67.

Mitchell, M., P. Hraber, and J. P. Crutchfield. "Revisiting the Edge of Chaos: Evolving Cellular Automata to Perform Computational Tasks." *Complex Systems* 7 (1993): 89–130.

Mitchell, Melanie. *An Introduction to Genetic Algorithms.* Cambridge, Mass.: MIT Press, 1996.

Moody, Glyn. "The Wild Bunch." *New Scientist* 160, no. 2164 (December 12, 1998): 42–46.

Nahapet, Janine and Sumantra Ghoshal. "Social Capital, Intellectual Capital, and the Organizational Advantage." *Academy of Management Review* 23, no. 2 (1998): 243–266.

Nelson, Richard and Sidney Winter. *An Evolutionary Theory of Economic Change.* Cambridge, Mass.: Harvard University Press, 1982.

Nesse, Randolph M. and George C. Williams. "Evolution and the Origins of Disease." *Scientific American* 279 (November 1998): 86–93.

Neustadt, Richard E. and Ernest R. May. *Thinking in Time: The Uses of History for Decision Makers.* New York: The Free Press, 1986.

New York Times Sunday Magazine Special Supplement, April 20, 1997.

Nisbett, Richard and Lee Ross. *Human Inference: Strategies and Shortcomings of Social Judgment.* Englewood Cliffs, N.J.: Prentice Hall, 1980.

Norman, Donald. Public lecture, School of Information, Ann Arbor, Mich., April 23, 1997.

Nye, Jr., Joseph S. and William A. Owens. "America's Information Age." *Foreign Affairs* 75, no. 2 (1996): 20–36.

Padgett, J. and C. Ansell, "Robust Action and the Rise of the Medici, 1400–1434." *American Journal of Sociology* 98 (May 1993): 1,259–1,319.

Page, Scott E. "On Incentives and Updating in Agent-Based Models." *Computational Economics* 10 (1997): 67–87.

Perrow, Charles. *Normal Accidents.* New York: Basic Books, 1984.

Pickard-Cambridge, Arthur. *The Dramatic Festivals of Athens.* Oxford: Clarendon Press, 1968.

Piore, Michael J. and Charles F. Sabel. *The Second Industrial Divide: Possibilities for Prosperity.* New York: Basic Books, 1984.

Pollack, Andrew. "The Judicial System vs. the Operating System." *New York Times* (July 20, 1998).

Poundstone, William. *The Recursive Universe.* Chicago: Contemporary Books, 1985.

Powell, Colin L. *My American Journey.* New York: Random House, 1995.

Putnam, Robert D. *Making Democracy Work: Civic Traditions in Modern Italy.* Princeton, N.J.: Princeton University Press, 1993a.

Putnam, Robert D. "Social Capital and Public Affairs." *The American Prospect,* no. 13 (1993b): 1–8.

Pyke, Frank, Giacomo Becattini, and Werner Sengenberger, editors. *Industrial Districts and Inter-firm Co-operation in Italy.* Geneva: International Institute of Labor Studies of the International Labor Organization, 1990.

Raymond, Eric S. "The Cathedral and the Bazaar." http://tuxedo.org/~esr/writings/cathedral-bazaar/cathedral-bazaar.html, August 8, 1998.

Reiter, Dan. *Crucible of Beliefs: Learning, Alliances, and World Wars.* Ithaca, N.Y.: Cornell University Press, 1996.

Resnick, Paul, Neophytos Iacovou, Mitesh Suchak, Peter Bergstrom, and John Riedl. "GroupLens: An Open Architecture for Collaborative Filtering of Netnews." In *Proceedings of ACM 1994 Conference on Computer Supported Cooperative Work,* 175–186. Chapel Hill, N.C.: Association for Computing Machinery, 1994.

Resnick, Paul. "Filtering Information on the Internet." *Scientific American* 276 (March 1997): 62–64.

Richardson, Lewis. *Statistics of Deadly Quarrels.* Pittsburgh: Boxwood Press, 1960.

Riolo, R. L. "The Effects and Evolution of Tag-Mediated Selection of Partners in Evolving Populations Playing the Iterated Prisoner's Dilemma." In *Proceedings of the International Conference on Genetic Algorithms (ICGA-97),* edited by Thomas Back, 378–385. San Francisco: Morgan Kaufmann, 1997.

Rosen, Stephen Peter. *Winning the Next War: Innovation and the Modern Military.* Ithaca, N.Y.: Cornell University Press, 1991.

Sampson, Robert J. et al. "Neighborhoods and Violent Crimes: A Multilevel Study of Collective Efficacy." *Science* 277 (August 15, 1997): 918–924.

Samuel, A. L. "Some Studies in Machine Learning Using the Game of Checkers." *IBM Journal of Research and Development* 3 (1959): 211–229.

Saxenian, Annalee. *Regional Advantage: Culture and Competition in Silicon Valley and Route 128.* Cambridge, Mass.: Harvard University Press, 1994.

Schelling, Thomas C. *Micromotives and Macrobehavior.* New York: W. W. Norton, 1978.

Schroeder, Manfred. *Fractals, Chaos, Power Laws.* New York: Freeman, 1991.

Shapiro, Carl and Hal R. Varian. *Information Rules: A Strategic Guide to the Network Economy.* Cambridge, Mass.: Harvard Business School Press, 1998.

Sharpe, William F. et al. *Investments.* Englewood Cliffs, N.J.: Prentice Hall, 6th ed., 1998.

Shardanand, U. and P. Maes. "Social Information Filtering: Algorithms for Automating 'Word of Mouth.'" In *Proceedings of the Conference on Human Factors in Computing Systems, CHI '95.* Denver, Colo.: Association for Computing Machinery, 1995, 210–217.

Shenitz, Bruce. "Review of Price of a Dream." *Smithsonian* 128 (September 1997): 115–117.

Simon, Herbert A. "A Mechanism for Social Selection and Successful Altruism." *Science* 250 (December 21, 1990): 1,665–1,668.

Simon, Herbert A. *Sciences of the Artificial.* Cambridge, Mass.: MIT Press, 2d. ed. 1981, 193–229.

Smith, Gaddis. *Morality, Reason, and Power: American Diplomacy in the Carter Years.* New York: Hill and Wang, 1986.

Sproul, Lee and Sara Kiesler. *Connections: New Ways of Working in the Networked Organization.* Cambridge, Mass.: MIT Press, 1992.

Stanton-Salazar, Ricardo D. "A Social Capital Framework for Understanding the Socialization of Racial Minorities Children and Youths." *Harvard Education Review* 67 (1997): 1–40.

Stinchcombe, A. L. and C. A. Heimer. *Organization Theory and Project Management: Administering Uncertainty in Norwegian Offshore Oil.* Oslo: Norwegian University Press, also Oxford: Oxford University Press, 1985.

Tanese, R. "Distributed Genetic Algorithms for Functional Optimization." Ph.D. thesis, University of Michigan, 1989.

Thompson, W. R. "On the Likelihood That One Unknown Problem Exceeds Another in View of the Evidence of Two Samples." *Biometrika* 25 (1933): 275–294.

Unger, Danny. *Building Social Capital in Thailand: Fibers, Finance, and Infrastructure.* Cambridge: Cambridge University Press, 1998.

Uzzi, Brian. "Social Structure and Competition in Interfirm Networks—The Paradox of Embeddedness." *Administrative Science Quarterly* 42, no. 1 (1997): 35–67.

Van Slyck, Abigail. *Free to All: The Carnegie Libraries and American Culture, 1890–1920.* Chicago: University of Chicago Press, 1995.

Waldrop, M. Mitchell. *Complexity: The Emerging Science at the Edge of Order and Chaos.* New York: Simon and Schuster, 1994.

Watts, Duncan J. and Steven H. Strogatz. "Collective Dynamics of 'Small-World' Networks." *Nature* 393, no. 4 (June 1998): 440–442.

Wilkinson, Lawrence. "How to Build Scenarios." *Wired* (September 1995, Scenarios: 1.01 Special Edition): 74–81.

Wilson, E. O. *Sociobiology: The New Synthesis.* Cambridge, Mass.: Harvard University Press, 1975.

Womack, James P., Daniel T. Jones, and Daniel Roos. *The Machine That Changed the World.* New York: HarperPerennial, 1990.

Young, Frank W. and Nina Glasgow. "Voluntary Participation and Health." *Research on Aging* 20 (1998): 339–361.

Yunas, Muhammad. "The Grameen Bank." *Scientific American* 281 (November 1999): 114–119.

Zimmerman, William and Robert Axelrod. "The 'Lessons' of Vietnam and Soviet Foreign Policy." *World Politics* 34 (1981): 1–24.

Index